67TH CONGRESS }
  3d Session }

HOUSE OF REPRESENTATIVES

I0060857

DEPARTMENT OF THE INTERIOR
ALBERT B. FALL, Secretary

UNITED STATES GEOLOGICAL SURVEY
GEORGE OTIS SMITH, Director

Bulletin 743

GEOLOGY OF THE OATMAN GOLD DISTRICT ARIZONA

A PRELIMINARY REPORT

BY

F. L. RANSOME

WASHINGTON
GOVERNMENT PRINTING OFFICE
1923

CONTENTS.

ILLUSTRATIONS.

GEOLOGY OF THE OATMAN GOLD DISTRICT, ARIZONA.

A PRELIMINARY REPORT.

By F. L. RANSOME.

INTRODUCTION.

The field work upon which the present report is based was begun December 18, 1920, and completed April 15, 1921. The district was geologically mapped in detail, and the workings of such mines as were in operation at that time were studied. In April, 1921, the only productive mines in the district were the Aztec mine of the Tom Reed Gold Mines Co. and the United Eastern mine, and both were believed by many supposedly well-informed people to be nearing the end of their principal known ore bodies. The Gold Road mine, formerly a close rival of the Tom Reed in productivity, was idle. A large body of ore was known to exist in the Big Jim mine of the United Eastern Mining Co., and some excellent ore had been found in the American mine. On the whole, the outlook for long-continued mining on an extensive scale can not be said to have been particularly bright at the time when the field work was finished. At the end of the year, however, two events led to greatly increased activity in the district. One was the finding of a small body of rich ore near the surface on the Stoney-Ferra lease block, on the Tom Reed vein; the other, still more significant, was the discovery of ore by means of the diamond drill on the 600 level of the Oatman United mine, where prospecting had been carried on with persistency long after most of the shafts sunk in the stirring days of 1915 had been abandoned.

The success of the Oatman United has led to renewed efforts to find additional ore bodies, particularly by diamond drilling, and consequently to demands for information concerning the geology of the district. It is to meet, if possible, this demand that the present preliminary report has been prepared. The reader should understand that whereas the statements contained within it are believed to be generally correct, nevertheless much of the material collected in

1

the course of the field work is still unstudied, and many problems are yet unsolved. The statements made and particularly such deductions or conclusions as are now presented are subject to revision in the final report. To that report will be deferred particular acknowledgments of courtesies extended and aid given, as well as full bibliographic references to previous publications on the Oatman district.

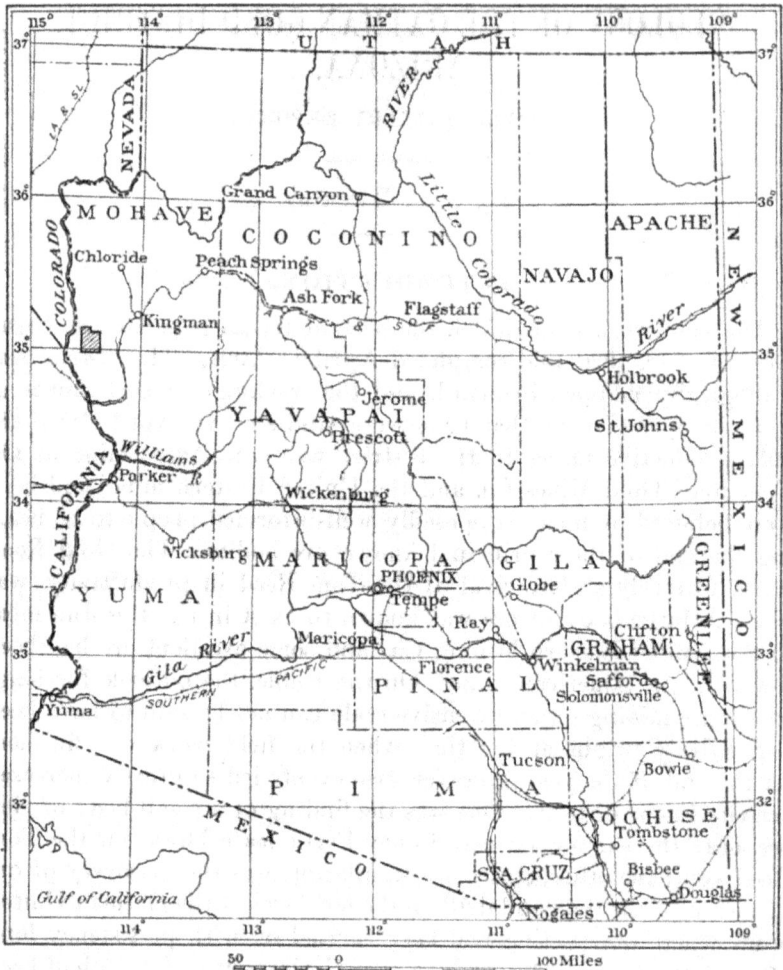

FIGURE 1.—Index map showing the situation of the Oatman district, Ariz. (shaded area).

No attempt is made in the present report to describe individual mines and prospects.

SITUATION.

The area described in this report as the Oatman district (fig. 1) lies in western Arizona, in Mohave County, mainly on the western slope of the Black Mountains, particularly that part of the mountains to which the names Black Mesa and Ute Mountains or Ute

Mesa have sometimes been applied. Oatman, the principal settlement, is 12 miles east of Colorado River and about 23 miles southwest of Kingman, the county seat, which is a thriving town on the main line of the Atchison, Topeka & Santa Fe Railway. An excellent road, 30 miles in length, connects Oatman with Kingman, crossing the Black Mountains through Sitgreaves Pass. In a straight line Oatman is 16 miles northeast of Needles, which is on the California side of the Colorado. Direct road connection with Needles is made, however, by way of the bridge at Topock, 16 miles by road southeast of Needles, where the Santa Fe Railway also crosses the river. A road has recently been completed between Oatman and Topock. A narrow-gage railway formerly connected the Leland and Vivian mines, in the Oatman district, with the east bank of Colorado River opposite Needles, but this railway, although it still appears on some maps, has long been abandoned, and the rails have been removed. Roads that are comparatively little used connect Oatman with a ferry near Needles, the Indian school at Mohave City on the Fort Mohave Indian Reservation, the old settlement of Hardyville, and other places on Colorado River. A rather rough road, with steep grades, runs north to Katherine and other points in the Union Pass district.

As used in the present report the name Oatman district designates only the southern part of the area comprised in what for purposes of record in the General Land Office is technically known as the San Francisco mining district. It does not include a number of mines to the north, among them the Thumb Butte, Arabian, and Katherine, which are also within the indefinitely bounded San Francisco district. The names Union Pass, Gold Road, Vivian, and Boundary Cone are sometimes used in connection with the word "district" to designate certain portions of the San Francisco district. They are merely local names that have no legal standing.

HISTORY OF MINING.

When, in 1862, Gen. J. H. Carleton and the Fifth California Volunteers occupied Arizona to prevent it from falling into the hands of Confederate irregulars, many of his men who were experienced miners varied the monotony of garrison duty by prospecting. At that time the main trail from Camp Mohave eastward ran up Silver Creek, and the first settlement in what soon became the San Francisco district was made on this stream at a point about 4 miles north of the site of the future town of Oatman. Here, where water can usually be obtained at the surface throughout the year, the walls of a dozen or more stone cabins, commonly reported to have been built by Carleton's men, are still standing.

About this time, in 1863 or 1864, John Moss found free gold in a large lode, now known as the Moss vein, which crops out prominently about a mile north of Silver Creek. Moss is commonly reported to have taken out rich ore to the value of $240,000 from a pocket close to the surface, but although considerable development work was afterward done on the vein, as a whole it proved to be of disappointingly low tenor. Some of the ore was worked in an arrastre on Silver Creek near the settlement. Afterward, it is said, a 10-stamp mill was built in the same locality, and still later a mill was erected on the Colorado near the mouth of Silver Creek, about 7 miles west of the mine.

The discovery of the Hardy vein, on the east side of Hardy Mountain; the Gold Dust (Victor-Virgin), about a mile southwest of Oatman; the Leland, on Leland Hill; and some others of the more prominently outcropping veins followed closely that of the Moss. Other very conspicuous veins, such as the Tom Reed and Gold Road, must have attracted the notice of the early prospectors and were probably located, although no ore was found in them until many years later.

For fully 30 years after the early work on the Moss and Hardy veins there is little record of activity in the Oatman district. It was not until 1901, when the Gold Road Co. sank the Tom Reed and Ben Harrison shafts to a depth of 100 feet on what is now generally known as the Tom Reed vein, that there was any general revival of mining.

About 1902 gold was discovered in the Gold Road vein. The ground was worked in a small way for several years under various owners, and at the time of the visit by F. C. Schrader, of the United States Geological Survey, in 1906-7,[1] it was owned by the Gold Road Mining & Explorations Co., incorporated under the laws of Arizona, with headquarters at Paris and Los Angeles.

In 1903 the Leland mine was bought by the Mount Mohave Gold Mining Co., afterward the Mohave Gold Mining Co. Without waiting to determine whether the ore bodies in the mine were such as to justify the expenditure involved, the new company built 17 miles of narrow-gage railway from the mine to a point on the Colorado opposite Needles and erected a mill at a place named Milltown, about 11 miles from the mine, on the sloping plain that intervenes between the Black Mountains and the river. Considerable ore was stoped from tunnels in Leland Hill and, according to Schrader, about 4,500 tons of ore was milled. The value of the total output from the mine is locally reported to have been about $40,000. About the end of 1904 the mine was closed, and at the time

---

[1] Schrader, F. C., Mineral deposits of the Cerbat Range, Black Mountains, and Grand Wash Cliffs, Mohave County, Ariz.; U. S. Geol. Survey Bull. 397, 1909.

of Schrader's visit many miles of the railway had already been washed out by floods.

In January, 1904, were located the Grey Eagle and Bald Eagle claims, which, as part of the Tom Reed ground, were to figure later in a notable lawsuit. In the same year the Tom Reed mine was sold, but failure of the purchasers to make payment led to the acquisition of the ground by the Title Insurance & Trust Co., of Los Angeles, as representative of various creditors. Shortly thereafter the Blue Ridge Gold Mines Co. was formed, took over the mine, built a 10-stamp mill, and began production. This company, however, was not successful, and in 1906 it sold the mine to the Tom Reed Gold Mines Co., the present owner. Up to April 1, 1922, the Tom Reed mines had produced bullion to the value of $9,488,622.

In 1911 the Gold Road mine was bought by the United States Smelting & Refining Co. at a price, as reported in the mining journals at the time, of $1,600,000.

Between the years 1903 and 1915 the history of the Oatman district was comparatively uneventful. The Tom Reed and Gold Road mines continued fairly steady production, prospecting was carried on in various parts of the district in a rather quiet way, and from time to time some of the smaller mines, such as the Ruth, Vivian, Pioneer (formerly the German-American), Gold Dust (formerly the Victor-Virgin), Midnight, and Gold Ore, yielded some ore.

Oatman, originally a very small settlement dependent upon the Tom Reed mine, appeared first under its present name in the mining news of the year 1912, when there seems to have been some renewal of activity in prospecting. During that year the Telluride Mining, Milling & Development Co. put up a hoist and began extensive underground exploration.

In March, 1913, J. L. McIver obtained an option on part of the ground now owned by the United Eastern Mining Co. No distinct vein was exposed at the surface, but the observant prospector noted that the rocks on the two sides of the suspected line of the vein were slightly different in character, and this suggested the existence of a fault fissure. In November, 1913, McIver and his partner, George W. Long, incorporated the United Eastern Mining Co., and in March, 1914, they began sinking a shaft at the north end of the town. After a depth of 40 feet had been attained work was stopped until additional money could be found. In June, 1914, W. K. Ridenour, noted locally for his enterprise and sagacity, came to the rescue of the new company by taking stock at 25 cents a share and recommending to his friends the purchase of additional shares. Up to December, 1914, about $13,000 had been raised by the sale of stock in comparatively small amounts. More money was necessary, however, to take up the option and to continue development. The late Frank A.

Keith happened about this time to be in Oatman, and, being convinced that McIver and Long had a reasonable chance of success, he brought the venture to the attention of Seeley W. Mudd and Philip Wiseman, well-known mining engineers of Los Angeles. Keith, Mudd, and Wiseman put up $5,000 each and were joined by R. I. Rogers and C. H. Palmer, jr. Altogether, $50,000 was paid for the present holdings of the United Eastern Mining Co., mainly in cash but partly in stock; the final payment was made June 15, 1916. In March, 1915, a crosscut on the 465 level went through 25 feet of ore that assayed $22.93. By the end of 1916, after an ore body estimated to contain gold to the value of about $6,000,000 had been blocked out, a 200-ton mill had been built and a new shaft in the hanging wall of the vein had been completed and equipped. The mill was afterward enlarged to a capacity of 300 tons. Up to the end of 1920 the United Eastern mine had yielded its fortunate owners 377,138 tons of ore of an average gross value of nearly $22 a ton, at an average operating cost of about $8.50. To the end of 1921 the company had paid dividends amounting to $3,952,700.

Such success in a district wherein practically no new ore bodies had been found for many years was naturally followed by a great revival of prospecting. Early in 1915 an exciting boom was in full swing at Oatman. The fact that the United Eastern ore body did not reach the surface—that it was found, in fact, where there was no evident sign of a vein at the surface—gave a color of reasonableness and an appearance of probability to ventures that ordinarily would have received no financial support. The commonly expressed belief that it was necessary to sink from 300 to 500 feet before an ore body could be looked for has been recorded in the scores of shafts which were sunk all over the district and most of which, in 1921, were abandoned. Although most of the work done during the period of excitement that followed the success of the United Eastern was futile, a few efforts were rewarded.

About 1914 the Tom Reed Co. had begun the development of its Black Eagle claim, at the extreme southeast end of its long strip of mining ground. Work on the 421-foot level of the Black Eagle shaft showed the existence of a vein which was considered sufficiently promising to invite further exploration, and in 1916 drifting was in progress on the vein at a depth of 565 feet, just above the level of underground water. About the same time what is now known to be the same vein was cut by the Big Jim Mining Co. in ground adjoining the Grey Eagle claim of the Tom Reed Co.

The Big Jim claim had been located on September 2, 1908, by S. S. Jones and others, not because it showed any indications of a vein but because it happened to be unlocated ground near a producing mine and might consequently have some speculative value. It lay immediately northeast of the Grey Eagle and Black Eagle claims, which cover a part of the Tom Reed vein lying southeast of what at that time was known as the productive part of the lode. Along this undeveloped part of the vein conspicuous outcrops are lacking, and this fact might well have been considered as giving additional chances of success to a claim located alongside of the Grey Eagle.

During the period of active prospecting that followed the discovery of the United Eastern ore body a shaft was sunk on the Big Jim close to the so-called discovery vein, an insignificant quartz-calcite stringer, of no economic importance. This vein was so obviously unpromising that no drifts were run on it. On the 400 level, however, a crosscut was run to the southwest, toward the supposed line of the Tom Reed vein, and at a distance of about 180 feet from the shaft this crosscut entered what has since been known as the Big Jim vein.

After the discovery of the Big Jim vein the Tom Reed Co. in 1916 sank the Grey Eagle and Aztec shafts. The Grey Eagle was intended particularly to investigate the relation of the Big Jim vein to the Tom Reed vein, and the Aztec was projected to give convenient access to the ore discovered from the Black Eagle shaft. Underground work from the Grey Eagle shaft soon showed that from 400 to 500 feet below the surface the vein is cut off by a fault that strikes nearly parallel with the vein. The vein dips northeast, and the fault fissure dips southwest. Ellis Mallery, then a young geologist employed by the Tom Reed Co., appears to have been the first to reach the conclusion, which development has since clearly shown to be true, that the vein in the Grey Eagle mine is the upper, downfaulted portion of the Big Jim vein. The principal displacement is along the fissure now appropriately known as the Mallery fault, but the present distribution of the ore bodies has been effected in part also by other faults, which will be described in another part of the report. The Grey Eagle work also resulted in the finding of profitable ore bodies in an intermediate segment of the vein between the Big Jim vein and what has been called the Grey Eagle or Tom Reed vein. (See fig. 6, p. 40.)

In April, 1917, the United Eastern Co. purchased the Big Jim mine. The able mining engineers who were directing the United Eastern Co. were well aware that the Big Jim and Grey Eagle veins were probably once continuous and owed their separation to faulting, but they were convinced that the amount of the throw, about 400 feet, was too great to enable the Tom Reed Co. to make successful claim to the apex of their vein. It soon appeared, however, that the Tom Reed Gold Mines Co. held a different opinion, and after an attempt to agree upon terms of consolidation had failed, this company, on February 17, 1919, filed suit against the United Eastern Mining Co.

The case was tried at Kingman, in the superior court of Mohave County, and was decided in favor of the United Eastern Mining Co. on March 28, 1921. Appeal was taken, and at the date of writing the case had not been finally decided.

In 1920 some excellent ore was found in the American mine, owned by the United American Mining Co., in the Tom Reed vein, southeast of the part worked in the Aztec mine. At the time of visit, early in 1922, this mine was still in the development stage.

The outstanding events in the early part of the year 1922 were the discovery of ore by diamond drilling from the 600 level of the Oatman United mine, about a quarter of a mile east of the town of Oatman, and the success of Stoney & Ferra, lessees, in finding rich ore near the surface in the Tom Reed vein, practically within the town. These successes have led to a renewal of prospecting on all sides, particularly by the use of diamond drills. The Tom Reed Co. has continued to work its Aztec mine on a diminished scale and has opened its mill to the ores of other mines, especially the American and Telluride, enabling these to carry on development work with some returns from ore already in sight. The United Eastern Mining Co., early in the year, completed an aerial tramway from the Big Jim mine to its mill near the No. 2 or main shaft of the United Eastern mine and began milling Big Jim ore. It also continued energetic prospecting, both by drilling from the workings of the Eastern mine and by sinking a new shaft to the north of its No. 2 shaft, to determine whether any additional ore bodies could be found in its original territory. In May it was reported that the Gold Road mill was being overhauled preparatory to the reopening of the Gold Road mine.[1a]

Recently an attempt has been made to work the gravels of Silver Creek for placer gold by sinking and drifting. Reports in the local press indicate that some gold has been found, but it is not yet certain that the metal is present in sufficient quantity to justify extensive operations.

TOPOGRAPHY.

Lying mostly on the western slope of the Black Mountains, the Oatman district as a whole inclines westward, toward the Colorado. The highest elevation attained within the area mapped in connection with this report is 4,955 feet above sea level, in the extreme northeast corner of the area, although Nutt Mountain, within the unmapped northeastern portion of the quadrangle covered by Plate I, reaches a height of 5,065 feet above sea level. The lowest ground

---

[1a] Since this report went to press high-grade ore has been mined from the Telluride vein by the Telluride and Tom Reed companies.

A. WESTERN FRONT, LOOKING NORTHEAST FROM A POINT NEAR OATMAN.

The original or No. 1 shaft of the United Eastern mine appears in the right foreground, at the north end of the town. The present main or No. 2 shaft, in the left center, is in the hanging wall of the vein. The dump in the left foreground is that of the Olla Oatman shaft, on the southwest branch of the Tom Reed vein. The flows whose edges appear along the mountain side are part of the Gold Road latite. The lighter-colored, irregular areas are intrusive rhyolite.

B. VIEW NORTHEAST OVER THE BROAD TOP OF THE MOUNTAINS
SOUTHEAST OF OATMAN.

Part of the western slope of the range appears on the left. The light-colored crag on the
extreme left is the rhyolite plug known as The Elephant's Tooth.

TOPOGRAPHY OF THE BLACK MOUNTAINS, NEAR OATMAN, ARIZ.

A. GENERAL VIEW TO THE NORTHWEST OVER THE CENTRAL AND NORTHWESTERN PART OF THE DISTRICT, FROM UTE MOUNTAIN.

The town of Oatman is mostly to the right of the view. The rough peaks in the foreground are Esperanza trachyte. The distant mountains are beyond the Colorado.

B. LELAND HILL FROM THE SOUTHEAST.

The shaft and mill low down in front of the hill are those of the Vivian mine. Higher and a little to the left is the Mitchell shaft of the Leland workings. The Leland vein is marked by a line of tunnels and dumps on the right or east slope of the hill. The prevailing rock in the view is the Alcyone trachyte.

TOPOGRAPHY OF THE OATMAN DISTRICT.

A. MOUNTAINS ON THE NORTHWEST SIDE OF TIMES GULCH.

The prevailing rock is Times porphyry.

B. THE RUTH AND MOSS MINES FROM THE SOUTH.

The prevailing rock is the Moss porphyry. In the left center are the shaft and mill of the Ruth mine. Above and a little to the left is the shaft of the Moss mine, with the prominent outcrops of the Moss vein stretching away to the right.

TOPOGRAPHY OF THE OATMAN DISTRICT.

A. HILLS CAPPED BY ESPERANZA TRACHYTE, 1½ MILES SOUTH OF OATMAN,
LOOKING SOUTHWEST.

Shows characteristic topography of the Esperanza trachyte with talus slopes of hard, clinking rock flakes. The White Chief shaft, in the left center, is in Alcyone trachyte. Directly behind it is an exposure of the sedimentary beds that separate the two trachytes.

B. BOUNDARY CONE FROM THE NORTHEAST.

A plug of nearly white rhyolite that has been thrust up through Alcyone trachyte.

TOPOGRAPHY OF THE OATMAN DISTRICT.

in the district is in the southwest corner, where the bottoms of some of the gullies are only about 1,450 feet above sea level. The range of relief within the area mapped on Plate I is accordingly close to 3,500 feet.

The district shows considerable diversity in topographic form, some parts being only moderately hilly, with comparatively smooth slopes, and other parts being characterized by slopes so steep and rough as to be impassable. Occupying from one-third to one-half of the width of the area, along the eastern border, is the principal ridge of the Black Mountains, divided, about halfway between the north and south boundaries of the district, by the east-west Sitgreaves Pass, 3,600 feet above the sea, through which runs the main road between Oatman and Kingman. These mountains have very clearly been carved from gently dipping volcanic rocks, chiefly lava flows, and have the general character of a dissected mesa. The degree of dissection, however, is far from uniform. North of Sitgreaves Pass the mountains, as a rule, have fairly sharp summits, from which the descent to the neighboring canyons is made by a series of great steps, of which the treads are gently sloping platforms, corresponding approximately to the upper surface of some lava flow, and the risers are cliffs, in places 300 to 400 feet in height. Battleship Mountain, of which only the south end comes within the area mapped, north and west of Cottonwood Canyon, is a conspicuous example of this type of topography.

South of Sitgreaves Pass for 2 miles, to the pass at the head of Antelope Canyon, the mountain slopes are rough and steep, but the bench and cliff topography is much less prominently displayed than farther north. The steep western front of this part of the range, as viewed from a point near Oatman, appears in Plate II, A.

South of Antelope Canyon the part of the Black Mountains within the area mapped, locally known in part as Ute Mountain, is decidedly mesa-like in character, with broad stretches of rolling upland (Pl. II, B) trenched by Warm Springs Canyon and its branches.

From the crest of the Black Mountains the descent is generally abrupt for about 1,000 feet to the hilly country in the vicinity of Oatman. A belt of such hilly country of very irregular width intervenes between the main ridge of the Black Mountains and the broad alluvial slope that stretches westward to the Colorado and is represented by a few square miles of country in the southwest corner of the area mapped on Plate I. Beyond the boundary of the mapped area this slope extends about 9 miles west of the river.

The general character of the topography of the hilly country between the main ridge of the Black Mountains and the valley of the Colorado is illustrated by Plate III, A. Within this belt the hills show much variety in form and a wide range in height. There are no long, even-crested ridges, such as result in many places where areas of folded or faulted sedimentary rocks are eroded, but the characteristic features of the topography are groups or clusters of pointed hills with no recognizable linear arrangement. Directly west of Oatman the hills, carved from andesitic and trachytic rocks, are comparatively low, rising for the most part less than 500 feet above the adjacent drainage channels or "washes." The summit of Leland Hill, the most conspicuous eminence in this part of the area, stands about 750 feet above the streamway shown in the foreground of Plate III, *B*.

Farther north, between Times Gulch and Silver Creek (Pl. I), is a group of higher and decidedly rugged hills, of which Mount Hardy (3,231 feet) is the dominant peak, and Cathedral Rock one of the most picturesque members. A view of these hills, which are composed partly of pink granite porphyry (Times porphyry) and partly of gray lavas (Alcyone trachyte), is shown in Plate IV, *A*.

In the northwestern part of the district is an area, 5 or 6 square miles in extent, underlain by monzonite porphyry—the rock of the Moss mine. Generally this rock gives rise to a comparatively low and open topography, with relatively small, sharp hills where the rock has been locally hardened by silicification or is traversed by dikes or veins. Northwest of the Moss mine, however, especially in the vicinity of the contact between the porphyry and the overlying volcanic rocks, some of the hills, as shown in Plate IV, *B*, are fairly high and steep.

In the southern part of the district, south of Oatman, a flow of compact, hard lava that rests on softer material has given rise to hills of the type shown in Plate V, *A*. The upper portions of these hills are as a rule extremely rugged, and the lower slopes are mantled with talus composed of clinking rock fragments.

This brief characterization of the hill forms of the Oatman district would be incomplete without mention of two peaks that command attention from every visitor on account of their light color and tower-like form. One of these is the well-known Boundary Cone, $3\frac{1}{2}$ miles southwest of Oatman, shown in Plate V, *B*. Why this mountain should have been named Boundary Cone is not clear, for, so far as known, it has never defined any important line of demarcation. The second peak is the small mass known as The Elephant's Tooth, which is a prominent feature of the eastern landscape as seen from Oatman and is shown in Plate II, *B*. As seen from the town this jutting mass of white rock has some resemblance to the fang of a carnivore but no real resemblance to any portion of the dental equipment of an elephant. The Elephant's Tooth and Boundary Cone are both plugs of rhyolite that, after being intruded into the surrounding rocks, have been stripped partly bare by erosion.

As the preceding description indicates, the Oatman district drains in general westward to the Colorado, although the water from the relatively small areas tributary to Meadow Creek, east of Sitgreaves Pass, and Antelope Canyon flows eastward into Sacramento Valley.

The principal drainageway is the channel of Silver Creek, which heads under Sitgreaves Pass and runs in a general northwesterly direction. Throughout most of the year the greater part of this channel is dry, but a small stream of water may persist through the summer north of Mount Hardy, near the site of the abandoned stone cabins of the California Volunteers. Excellent perennial water is found also in the upper branches of Cottonwood Creek, which is a northerly affluent of Silver Creek. This water, which issues from springs, supplies the Tom Reed mine and the town of Oatman.

A broad, flaring drainage channel, with many tributary washes, such as Iowa and Esperanza canyons from the south and Leland and Rattlesnake gulches from the north, heads near Oatman and merges to the southwest with the alluvial slope of the Colorado. This channel and its tributaries are generally waterless.

Times Gulch, which drains the west-central part of the district, is also normally dry, although it evidently on occasions carries a large volume of flood water.

GENERAL GEOLOGY.

OUTLINE.

The rocks of the Oatman district comprise a closely related series of Tertiary volcanic flows, with associated tuffs and some subordinate beds of conglomerate or breccia, sandstone, shale, and limestone. The volcanic flows rest as a whole on pre-Cambrian crystalline rocks, chiefly granitic, and have been invaded and displaced in part by masses of porphyry that are closely related in composition to some of the flows and probably came from the same bodies of molten rock material or magma. They are also cut by numerous dikes and small irregular intrusive bodies of rhyolite and other rocks.

The igneous rocks have a wide range in composition, from siliceous potassic rhyolite to olivine basalt, but the preponderant members of the series belong within the compositional range represented on the one hand by latitic andesite and, passing through latite, on the other hand by latitic trachyte. The discrimination and classification of rocks which are so closely akin and between which there are no very definite distinctions present considerable difficulty, particularly as lavas of practically identical composition were erupted at different times. Classification and nomenclature will be fully discussed in the final report and illustrated by chemical analyses of the principal rock types. It must suffice here to give merely the conclusions from chemical and petrographical studies, so far as these have at present gone.

The volcanic flows as a whole dip to the east. The average dip can not be determined satisfactorily, as some of the flows are thick and only here and there afford any indication of their attitude. Such observations as have been made, however, indicate that the general dip of the series is about 15°, although in places the inclination is much steeper. The flows of basalt, the latest lava to be erupted, are apparently more nearly horizontal than some of the older rocks. The prevailing easterly dip and the fact that the eastern part of the district (Pl. I) contains the crest of the Black Mountains result in the exposure of north-south belts of successively younger formations from east to west across the district. The pre-Cambrian rocks are found along the comparatively low western border of the area, and the latest basalt flows cap the high hills along the eastern border. In general, these hills show precipitous or steplike western fronts, with rough and approximately horizontal banding indicative of the accumulation of successive lava flows, but less abrupt eastern slopes, on which the edges of the flows are as a rule not visible.

The general structure of a simple earth block tilted gently to the east is modified to some extent by faulting and by the displacement that accompanied the intrusion of two large masses of porphyry. The older mass, extending from Silver Creek northward past the Moss and Mossback mines and beyond the boundary of the area mapped in Plate I, is quartz monzonite porphyry. It is here named the Moss porphyry, as the Moss mine is wholly within this rock. The younger mass is exposed within the triangular area bounded on the northeast by Silver Creek, on the southeast by Times Gulch, and on the west by the alluvial plain of the Colorado. This rock, a granite porphyry, forms the main mass of Hardy Mountain, the pinnacle known as Cathedral Rock, and the precipitous peaks with their characteristic faintly rosy hue that are so conspicuous on the north side of Times Gulch. It is here named the Times porphyry.

Much smaller but very conspicuous intrusive masses are the white rhyolite plugs of Boundary Cone and of The Elephant's Tooth, half a mile east of Oatman (Pls. II, B, and V, B).

The general sequence of events as recorded in the rocks of the district was as follows:

In Tertiary time, before volcanic activity began, the region in which the Oatman district lies was floored by deeply eroded pre-

Cambrian crystalline rocks. Prominent among these were dark medium-grained, more or less squeezed and foliated granitic rocks, probably to be classed as quartz-mica diorites, micaceous schists of various kinds, probably some true granite, and dike rocks of considerable diversity. At the present time these ancient rocks are exposed only in a few small areas along the western border of the district and where so visible are generally decomposed and crumbling. They may be seen around the abandoned Chicago Syndicate shaft, south of the mouth of Times Gulch, and near the Murdock shaft, southwest of Boundary Cone.

The erosion surface of early Tertiary time was probably more or less hilly, but no definite information as to its topographic character has been obtained.

At a time which can not be closely fixed but which from our general knowledge of the geology of Arizona may be considered as probably middle or late Tertiary, volcanic eruptions broke through the pre-Cambrian rocks. This event must have been accompanied and perhaps immediately preceded by changes in the surface configuration of the country and consequently in erosion and in the transportation and deposition of detritus. At some places, particularly near the mouth of Times Gulch, there is a coarse breccia, consisting of blocks of pre-Cambrian rock in a greenish matrix that appears to have been originally a mud of fine volcanic detritus. With the breccia are lenticular layers of greenish sandstone and a few thin and nonpersistent beds of shale. This deposit is interpreted as representing the rather rapid accumulation in local basins, at the beginning of the volcanic eruptions, of material that was partly the rock detritus of the old prevolcanic surface and partly fine volcanic particles blown from some of the newly opened vents and carried to their resting place by the disturbed and shifting streams of this period of transition. The maximum observed thickness of this deposit is estimated to be between 100 and 200 feet.

Overlying the breccia and in places resting directly upon the pre-Cambrian rocks is a series of flows with probably some intrusive masses, which in this report will be designated the Alcyone trachyte (al-sy'oh-nee), from the Alcyone mine. This formation includes what Schrader called the "basal andesite" or, in some parts of his report, the "older andesite." [2] It also includes, north of Times Gulch and west of Meals Camp, what Schrader referred to as "gray andesite." [3] The thickness of this formation is not known. It is roughly estimated, from the section across the district shown on Plate I, as about 2,800 feet, but this estimate, like other estimates

[2] Op. cit., pp. 34, 181.
[3] Op. cit., p. 174.

of the thickness of the principal volcanic formations, depends fundamentally upon the assumed dip of the formation as a whole. The figure given is based on a supposed easterly dip of about 15°. If the formation as a whole is more nearly horizontal, the maximum thickness may not be more than 1,000 feet. Schrader estimated its thickness as " at least several hundred feet." A part of the Alcyone trachyte, particularly in the extreme western part of the district, west of Oatman, may be intrusive into flows of essentially the same composition. If so, it is highly probable that at more than one place in the district the trachyte, instead of merely resting on the pre-Cambrian rocks, may extend indefinitely downward to what was originally its magmatic source.

The eruption of the Alcyone trachyte was followed by a brief interval of erosion, during which there was local accumulation of some sandy beds whose materials were apparently derived from the trachyte. These beds, with a thickness of about 115 feet, are exposed in a steep north slope about $1\frac{1}{2}$ miles south of Oatman and a quarter of a mile south of the abandoned White Chief shaft. (See Pl. V, A.) This exposure is visible from the Oatman-Topock road and rather conspicuous from the contrast between the red color of the upper layers and the dark shade of a lava cap above them. The same beds, here apparently thinner, are exposed in the saddle east of Boundary Cone and appear in the dump of the United Oatman tunnel, on the south side of Esperanza Canyon.

These sedimentary beds are overlain by a flow or flows of a very compact ringing lava which Schrader described as " a dense dark reddish-brown or purple andesite, locally known as ' phonolite ' from its metallic ring when struck." [4] Its local designation by those prospecting within it in 1921 was " latite," and such the rock might be called. As will be shown, however, in the final report, the rock is not a typical latite but rather a latitic trachyte, almost identical in composition with the Alcyone trachyte, although it is plainly younger than that rock and is very different in appearance. It will be named in this report the Esperanza trachyte, from the canyon of that name. The Esperanza trachyte at present occupies a comparatively small part of the district, extending from a point about $1\frac{1}{2}$ miles south of Oatman southward beyond the boundary of the area mapped, with its widest part just east of Boundary Cone. It is the prevailing rock of Iowa and Esperanza canyons and is the country rock of the Highland Chief mine. The maximum thickness of the Esperanza trachyte is estimated to be about 1,000 feet.

The next rock to be erupted after the Esperanza trachyte was the Oatman andesite, so named from the town of Oatman. This is

---

[4] Op. cit., p. 199.

what Schrader called "green chloritic andesite," a good descriptive designation but one that is rather cumbrous and not in accord with the present practice of the United States Geological Survey in giving rock names. This is the rock that has contained all the large ore bodies hitherto worked in the district with the exception of those of the Gold Road vein. As is more fully shown elsewhere in this report (p. 21), it is mainly a series of flows but in a number of places is clearly intrusive in the Alcyone trachyte, which appears generally to underlie it, separated by a thin accumulation of tuffaceous shaly and calcareous beds. Where the Esperanza trachyte is present, that rock, of course, intervenes between the Oatman andesite above and the Alcyone trachyte below. Owing to the partly intrusive character of the Oatman andesite and the probability that at various places within the district masses of it extend indefinitely and irregularly downward through the older rocks, the thickness of the formation at any particular place is wholly indeterminable. The general maximum thickness of the formation as a whole, without regard to such possible downward extensions, is roughly estimated to be about 2,700 feet. This estimate, however, like that of the Alcyone trachyte, depends upon the dip assumed. If the formation as a whole is approximately horizontal, the maximum thickness may not be more than 700 or 800 feet. Schrader's estimate was 800 feet.[5] On the other hand, since Schrader's examination the Tom Reed and United Eastern mines have attained a depth of approximately 1,500 feet without, so far as known, getting out of the Oatman andesite.

Directly overlying the Oatman andesite to the east and forming the greater part of the Black Mountains east of Oatman is an assemblage of rocks, chiefly volcanic flows but with some tuff and intrusive material, which is here provisionally named the Gold Road latite, after the settlement and mine of that name. This formation corresponds in a general way to Schrader's "undifferentiated volcanic rocks"[6] but probably includes also what he termed the "rhyolite series" and "younger andesite." It consists mainly of biotitic latite, but there have been mapped with it on Plate I some overlying glassy and lithophysal rocks which should perhaps be classed as rhyolite and which will be separately mapped and named in the final report. Between some of the flows are layers of tuff. The maximum thickness of this formation, which as mapped in this preliminary report includes all the flows exposed along the precipitous mountain front east of the town of Oatman, is estimated to be from 3,000 to 4,000 feet.

Overlying the Gold Road latite and well exposed in cliffs along Cottonwood Canyon, in the northeastern part of the area, is a flow of compact brown lava that closely resembles the Esperanza trachyte.

[5] Op. cit., p. 35.
[6] Op. cit., p. 37.

It is here named the Flag Spring trachyte, from Flag Spring, in Cottonwood Canyon. It forms a well-defined reddish-brown cliff on the south and east sides of Battleship Mountain and also the cliff just north of the Gold Ore mine. It apparently was not an extensive flow and is not represented south of Sitgreaves Pass. It attains a thickness of about 250 feet.

Succeeding the Flag Spring trachyte in age and overlapping it slightly in position is a flow of nearly identical rock, which, from its occurrence on the north side of Meadow Creek, is here named the Meadow Creek trachyte. This flow, like the Flag Spring trachyte, is not extensive and is not known south of Meadow Creek, up which runs the road from Kingman to Sitgreaves Pass.

Resting partly on the Gold Road latite, partly on the Flag Spring trachyte, and partly on the Meadow Creek trachyte is an accumulation of pumiceous cream-colored tuff which forms very conspicuous bluffs along Meadow Creek, on the north side of the Oatman-Kingman road. This is named the Sitgreaves tuff, from Sitgreaves Pass. Variable in thickness and resting in places against steep slopes, it was evidently deposited after the underlying lavas had been rather energetically cut into by erosion. It appears to have a maximum thickness of 200 to 300 feet.

Overlying the Sitgreaves tuff and forming the top of Battleship Mountain and of other peaks and ridges to the north and east is a flow of very glassy spherulitic lava, which in places is probably from 500 to 600 feet thick. It is found only in the northeastern part of the district, about the headwaters of Cottonwood Canyon, but extends for an unknown distance north and east of the area mapped. It is here named the Cottonwood rhyolite.

Finally, overlying all the rocks named and, south of Sitgreaves Pass, resting on the Gold Road latite, are thin flows of black olivine basalt, with associated red or gray tuffs. Undoubtedly the basalt once covered most of the range, but it has been reduced in area by erosion and now occurs as a capping on the higher hills and mesas. The most extensive exposure is that in the southeast corner of the district, where Ute Mesa, for many miles south and east of the area mapped, is covered by this somber rock. A considerable area is preserved also in the northeastern part of the area, north of Meadow Creek and east of Cottonwood Creek. The maximum thickness of the basalt and associated tuffs preserved within the area of the district as mapped in Plate I is about 1,000 feet.

After the eruption of the basalt the rocks of the region were probably faulted and tilted, and active erosion set in and has continued to the present time. While the higher slopes underwent continuous erosion during this period, erosion and deposition alternated in some of the lower portions of the area, as shown by the presence of alluvial terraces and gravel-capped ridges and by the dissection of some well-bedded fluvial deposits along Silver Creek.

DISTRIBUTION AND CHARACTER OF THE PRINCIPAL IGNEOUS FORMATIONS.

ALCYONE TRACHYTE.

As would be expected of the lowest member of a generally eastward-dipping series of lava flows, the Alcyone trachyte is found mainly in the western part of the district, where it is generally exposed over an irregular north-south belt as much as 2 miles wide. The group of high hills in the northwest corner of the district, west of Meals Camp and the Moss mine, are carved almost wholly from this rock. Together with the Times porphyry, which is intrusive into it, the Alcyone trachyte occupies the triangular area northwest of Times Gulch and southwest of Silver Creek. It is the prevalent rock from the west base of Leland Hill westward to the edge of the sloping detrital plain of the Colorado. It is the country rock of the Alcyone mine and is exposed on both sides of Oatman Wash as far east as the village of Old Trails. It is the rock through which protrudes the imposing rhyolite plug of Boundary Cone, and it extends southward beyond the limits of the mapped area.

From the northern boundary of the mapped area southward to Times Gulch the Alcyone trachyte is generally of uniform character. It is a gray vitrophyre—that is, a rock in which small distinct crystals, chiefly white feldspar, lie in a blotchy, streaky matrix or groundmass that is mainly volcanic glass crowded with microscopic crystals. This is a common texture in rather siliceous lavas that become viscous or pasty before they completely crystallize and continue to flow in such imperfectly liquid condition. The microscope shows the feldspar phenocrysts as sanidine, or potash feldspar, and andesine, or soda-lime feldspar. With these is usually a little biotite or black mica.

A chemical analysis of a typical specimen (O–142) of the Alcyone trachyte from the summit of a hill three-quarters of a mile westnorthwest of Cathedral Rock, made by R. C. Wells in the Survey laboratory, is as follows:

Chemical analysis of Alcyone trachyte.

SiO_2	65.26	H_2O above 110° C	0.34
Al_2O_3	16.39	H_2O below 110° C	.47
Fe_2O_3	1.98	TiO_2	.55
FeO	1.21	P_2O_5	.14
MgO	.83	MnO	.05
CaO	2.16	CO_2	.14
Na_2O	4.23		
K_2O	6.30		100.05

The rock contains more plagioclase and consequently more lime and soda than a typical trachyte and is close to the indefinite line that separates trachytes from latites. It might equally well be called a trachytic latite or a latitic trachyte, but as there are other latites in the area there is a gain in convenience and distinction in calling it a trachyte.

The line of Times Gulch marks in general a change in the appearance of the Alcyone trachyte which appears to be a gradation. The rock south of Times Gulch shows more alteration, and its general color changes from brownish gray to greenish gray. The vitrophyric texture is less evident, and much of the rock is a rather massive porphyry, which crumbles on weathering. There is evidence in places that this porphyry is intrusive into finer-textured varieties, but an attempt to map it separately had finally to be abandoned. In places, especially on some of the higher hills between Times Gulch and Oatman Wash, are flows that are darker and rather fresher than the formation as a whole but are not separable units. The formation in this area is apparently a volcanic complex in which flows were cut and invaded by intrusive bodies of the same magma, the whole belonging to one period of igneous activity.

Along its eastern border the Alcyone trachyte is generally overlain by the Oatman andesite, which in some places was poured over the trachyte but in other places has clearly been forced up in molten condition through it. Where the Oatman andesite overlies the trachyte the two rocks are generally separated by a deposit of sedimentary material of variable thickness and of considerable diversity in appearance. One of the most characteristic varieties is a rather massive shale which is pale or greenish gray as seen underground but which in weathered exposures is pale buff or cream-colored and is more obviously fissile than the unweathered rock. This shale is well exposed on the south side of Times Gulch, opposite the mouth of the gulch in which is the Oatman Belle shaft, where it is from 100 to 150 feet thick and might be mistaken for a compact rhyolite. It is also exposed at a number of places near the Midnight and North Star mines. In some places, as about a third of a mile northwest of Leland Hill, the creamy shale is overlain by gray calcareous shale with thin lenses of limestone. These shaly beds, although in many places too thin to map on the scale used for the geologic map of the Oatman district, appear to be fairly persistent and, if the mines near Oatman ever pass through the bottom of the Oatman andesite, should be carefully looked for as an indication that the Alcyone trachyte is not far below. The Gold Key shaft, about two-thirds of a mile northwest of Old Trails and 338 feet deep, shows some of the shale in the dump, and according to Mr. Charles T. Weidman the shaft passed out of the Oatman andesite into this material at 300 feet.

The shale has been described as resting on the Alcyone trachyte, but the exact boundary between the trachtye and the overlying sedimentary material is in many localities very obscure. The trachyte in its upper part contains many fragments, also of trachyte, and appears to be a flow breccia—that is, a lava which, after partly solidifying, continued to flow, breaking up its congealed crust and mingling the fragments with the still liquid lava. Such a rock can not always be distinguished from a sedimentary rock composed mainly of fragments of a lava upon which it rests or from rocks composed of particles of lava blown explosively from some volcanic vent. The difficulty in drawing a boundary between the Alcyone trachyte and the overlying sedimentary beds is well exemplified in the vicinity of the North Star mine. In many places, as just west of the Pioneer mine or between the Alcyone mine and Old Trails, the weathered trachyte exhibits rather conspicuous lamination, as shown in Plate VI. This is probably due to an original flow structure in the trachyte, which has been accentuated by weathering.

Near Boundary Cone and south of it the Alcyone trachyte contains some masses of volcanic rock that are more calcic than the trachyte and should be classed as latite or andesite. These appear to be intrusive in the trachyte as dikes or small irregular masses. It was not practicable to map them separately.

At the head of Times Gulch, east of Hardy Mountain and extending northeastward to Silver Creek, is an area of generally decomposed or altered rock of doubtful affinities. It has been provisionally mapped as Alcyone trachyte but is more biotitic than the typical trachyte and closely resembles some of the flows in the formation named Gold Road latite. As it appears, however, to underlie the Oatman andesite, which is certainly overlain by the Gold Road latite, this rock is mapped with the Alcyone trachyte.

In the northern part of the district, particularly north of Times Gulch, the Alcyone trachyte contains some lenticular inclusions of breccia, composed almost entirely of fragments of pre-Cambrian rock and practically identical with the breccia at the base of the volcanic series, exposed at the mouth of Times Gulch. Presumably, while the trachyte was being erupted some of the pre-Cambrian rocks were not covered by lava and, shattered possibly by volcanic explosions, furnished abundant angular detritus to be washed into local and transient basins and buried under later flows. Some of the inclusions are composed entirely of fresh granitic detritus cemented into a hard rock that looks and weathers so much like granite as to be exceedingly deceptive.

ESPERANZA TRACHYTE.

The Esperanza trachyte, a rock of distinctive character, is restricted to the southern part of the district and is typically exposed in Iowa and Esperanza canyons and in the vicinity of the Highland Chief mine.

The general color of the rock as seen in extensive exposures is reddish brown, but on freshly broken surfaces it is a rather light purplish gray. The texture is very compact, the only minerals recognizable to the unaided eye being black mica in minute sparkling foils and hornblende in equally small crystals. Occasionally a few small phenocrysts of feldspar, 4 millimeters or less in length, can be detected. The rock breaks with splintery fracture and is a decided cliff maker. Typical topography associated with it is shown in Plate V, A. Sweeping up to the faces of the cliffs are steep detrital slopes composed of hard ringing flakes of the trachyte. So sonorous are these fragments that the rock has sometimes been referred to as phonolite. It is not uncommon to find in the talus slabs from half to three-quarters of an inch thick and 18 inches across that ring when struck like a plate of steel and require a severe blow to break them. With the slabs are slender splinters, perhaps three-quarters of an inch thick, 2 inches wide, and 2 feet long.

Under the microscope the rock shows little of interest. It is mainly a very fine textured felt of tiny, obscure feldspar crystals (microlites) of low refractive index, with probably some glass. Although very different in its lithologic properties it is chemically nearly identical with the Alcyone trachyte. It contains, however, a little less potassium and a little more iron.

Most of the Esperanza trachyte is a massive lava flow, or possibly two or more such flows. Associated with the massive rock, however, are some irregular and rather indefinitely bounded masses of breccia, made up of fragments of the same rock. Whether these are flow breccias or the products of volcanic explosion could not be definitely determined. They are believed to be the products of local explosions and may mark approximately places where the trachyte broke through from below. The breccia has been cut by the massive trachyte.

The Esperanza trachyte extends for an unknown distance south of the area mapped, has a width of 2 miles east of Boundary Cone, and thence narrows rather rapidly northward, presumably because the flow decreases in thickness in that direction. Just north of Sunnyside Wash the trachyte rather abruptly disappears. The thinning northern portion of the flow is probably faulted down here, on the north side of the Gold Dust–Boundary Cone fault zone, and is covered to the north by the Oatman andesite.

OATMAN ANDESITE.

The Oatman andesite, the "green chloritic andesite" of Schrader and the rock in which the largest ore bodies in the district have been found, is exposed over a very irregular area. South of Sunnyside Wash it crops out as a comparatively narrow belt, from a quarter to half a mile wide, separated from the Esperanza trachyte below by some light-colored tuffaceous sandstone that has an estimated maximum thickness of 50 feet and overlain directly by the assemblage of lava flows that has been designated the Gold Road latite. North of Sunnyside Wash and, still more noticeably, north of Oatman Wash the area of Oatman andesite expands abruptly to a width of nearly 3 miles in the latitude of Oatman. At many places along the western border of this area the Oatman andesite rests upon the Alcyone trachyte, generally with a thin layer of shale at its base. This relation can be well seen southwest of the Pioneer mine and west and south of Leland Hill. In other places, as about a mile northwest of Leland Hill, the andesite is clearly intrusive into the trachyte and associated shale.

North of Oatman, along Silver Creek, the broad belt of Oatman andesite terminates abruptly, its place northward being taken by the intrusive Times porphyry and the Gold Road latite. The ending appears to coincide with a zone of faulting, which is obscure near the Oatman Amalgamated and Sun Dial mines but is supposed to be continuous on the northwest with the Gaddis & Perry lode and on the southeast with the Gold Road vein. If, as is probable, the downthrow along this fault is on its north side, the effect would be to shift to the west the outcrop of the Oatman andesite north of the fault. It is quite possible that the area mapped as Moss porphyry between the two main branches of Silver Creek may include some Oatman andesite. The rock of this particular area has been greatly altered, much of it beyond identification, by the introduction of finely disseminated pyrite and the subsequent oxidation of the sulphide with the generation of sulphuric acid. The iron of the original ferromagnesian minerals has been deposited as hydrous oxide, and the aluminous minerals have also been decomposed with the production of alum and other less soluble sulphates. Farther north the Oatman andesite is recognizable again at the Mossback mine and beyond the northern limits of the area mapped.

The Oatman andesite, although it comprises a number of flows and some intrusive rock, is as a whole fairly uniform in character and not difficult to identify in the field. As a rule it is more decidedly porphyritic than the rocks previously described, the larger crystals or phenocrysts of feldspar being particularly conspicuous. These are of moderate size, generally less than a centimeter in length, and contrast with the general dark color of the rest of the rock. The hue of the rock as a whole ranges from dark gray or nearly black in fresh specimens to moderately dark greenish gray in weathered or altered varieties. Some tint of green is almost invariably noticeable. With the exception of the basalt which caps some of the higher hills, the Oatman andesite has generally a more "basic" appearance than the other volcanic formations in the district. Unlike the trachytes it is, in many places, amygdaloidal—that is, it contains steam bubbles that have been filled with calcite, quartz, chlorite, or minerals belonging to the group known as zeolites. These filled bubbles, or amygdules, are particularly abundant in surface exposures south of Sunnyside Wash and near the Gold Dust mine. Some as large as 2 inches in diameter were noted. They were observed also underground in a number of mines.

At a few places in the Oatman andesite are found lenticular layers of greenish-gray tuffaceous sandstone. As this rock is rather soft and crumbling it is not as a rule well exposed at the surface. Some of it may be seen in the area of Oatman andesite southeast of the American mine and southwest of the Argo mine, where it appears to be about 25 feet thick. Another lens of about 20 feet in maximum thickness is exposed alongside the old road three-quarters of a mile west of Gold Road, where the sandstone and the Oatman andesite are cut by a small dikelike body of rhyolite. Mr. Oscar H. Hershey, who is familiar with the old workings of the Tom Reed mine, most of which could not be seen in 1921, informed me that in the course of his underground work he noted lenses of tuffaceous sandstone at various places in the Oatman andesite. The sandstone undoubtedly occurs between successive flows, perhaps at more than one geologic horizon, and represents intervals of erosion and local deposition of sandy detritus derived from the rocks previously erupted.

As shown by the microscope, the Oatman andesite consists essentially of feldspar (which ranges in composition from andesine to labradorite), augite, biotite, apatite, and magnetite in a groundmass of glass crowded with small crystals of feldspar and one or more of the other minerals named. In the freshest dark specimens the feldspar is clear, but as a rule it is partly decomposed and consequently milk-white and turbid. The augite in all but the freshest varieties of the rock has been changed to calcite. The biotite is very rarely fresh but as a rule has lost its dark-brown color as seen in thin sections and become pale green. The alteration product is, in a few specimens, identifiable definitely as aggregates of the mineral chlorite, but more commonly it seems to be a homogeneous intermediate product that retains some of the optical properties of the biotite. It is this prevalent alteration of the biotite which gives the Oatman andesite its greenish tint. In contrast with some of the flows of the Gold Road latite, the biotite of the Oatman andesite is never conspicuous, and neither it nor its alteration product can as a rule be recognized with the unaided eye.

An analysis of a sample of Oatman andesite obtained near the No. 1 shaft of the United Eastern mine, made by R. C. Wells in the Survey laboratory, is as follows:

Chemical analysis of Oatman andesite.

SiO_2	56.37	H_2O below 110° C	0.38
Al_2O_3	15.99	H_2O above 110° C	1.66
Fe_2O_3	2.56	TiO_2	1.16
FeO	3.41	P_2O_5	.41
MgO	2.97	MnO	.07
CaO	6.81	CO_2	2.18
Na_2O	2.99		
K_2O	2.89		99.85

This analysis shows that the Oatman andesite is less calcic and slightly richer in alkalies, especially in potash, than a typical andesite and approaches a latite in composition. Another sample analyzed, from Leland Hill, contains 2.68 per cent of soda and 4.13 per cent of potash. This particular variety might be classed as andesitic latite. Schrader[7] records that a sample collected by him from Leland Hill was even higher in potash, containing 2.60 per cent of Na_2O and 5.19 per cent of K_2O. He classified this rock, naturally enough, as latite, although he recognized it as being a part of his "green chloritic andesite."

Considered by itself the Oatman andesite might perhaps properly be called the Oatman latite. But it is notably richer in lime and magnesia than the other rocks of the district classed as latite and is also lower in alkalies. This difference has led to the choice of andesite rather than latite as a general petrographic designation of the formation.

GOLD ROAD LATITE.

The Gold Road latite is a volcanic complex of lava flows and tuffs, with a minor proportion of intrusive rock that appears to have been forced through some of the lower flows and to pass with no definite boundary into some of the upper flows. The flows present considerable variety in texture and composition, ranging upward from fairly typical almost wholly crystallized latites to black obsidian-like, pumiceous, or lithophysal flows that might be classed as glassy rhyolite and will be separately mapped and named in the final report.

In the southern part of the district the Gold Road latite is largely covered by basalt, only a narrow strip being exposed in the cliffs of

---

[7] Op. cit., p. 87.

Ute Mesa, at the heads of Iowa and Esperanza canyons. From the town of Oatman eastward to the boundary of the area studied the rocks forming the upper part of the main ridge of the Black Mountains belong, with few minor exceptions, to this formation. Northward, the Gold Road latite continues as an irregular belt past the Mossback mine and passes beneath the younger volcanic formations of the northeast corner of the district.

The basal flow of the Gold Road latite, particularly from Sunnyside Wash northward beyond the Mossback mine, is characterized by rather abundant sparkling flakes of biotite and somewhat larger phenocrysts of feldspar than the Oatman andesite. As a rule the latite is lighter in color than the Oatman andesite and does not have the green tint of that rock. A large proportion of the specimens from various localities show fairly conspicuous crystals of feldspar and many flakes of brilliant black mica in a dense pasty-looking groundmass that is pale red-gray or lilac-gray.

The actual contact between the Oatman andesite and the Gold Road latite is seldom well exposed. There is apparently no continuous separating layer of sedimentary material between the Oatman andesite and the latite, but in some places the andesite near the contact appears to be soft and decomposed, and in others it is a breccia, such as might occur on the upper surface of a lava flow. Apparently there was no time for much erosion between the eruption of the two lavas. The main criterion for mapping the contact is the difference in character of the upper and lower flows. After some local field experience the distinction between the latite with its abundant flashing mica flakes and the dark-green Oatman andesite, in which fresh crystals of biotite are rare and inconspicuous, can be made with certainty. This distinction was recognized by Hershey in his work for the Tom Reed Co. and was utilized in his preparation of the case for that company against the United Eastern Mining Co.

One locality where the distinction between the Oatman andesite and the Gold Road latite is fairly plain is about a mile south of Oatman, just west of the saddle through which passes the road to the American and Argo mines. The upper part of the hill west of this pass, like that of the similar precipitous hill above the American shaft, is composed of Gold Road latite. On the west side of this hill the contact between the Oatman andesite and the biotitic basal flow of the latite can be traced from the north edge of Sunnyside Wash, just north of the Pictured Rock shaft, northward through a saddle, and then down the northern slope of the hill to a point near the Telluride No. 3 shaft. West of this contact is a similar but lower hill also capped by the Gold Road latite, which rests on the Oatman andesite with a fairly well defined contact. The difference between the two rocks can also be well seen near the water tanks on the hillside above the Red Cloud shaft, north of Oatman, and the general line of the contact can be traced deviously around the hillsides past the Arizona Rex shaft to the vicinity of Gold Road. Along the ravine halfway between the Arizona Rex shaft and Gold Road, up which a road runs to what is supposed to be the United Northern shaft, the contact is displaced by what is thought to be the Mallery fault or a fissure of the Mallery fault zone.

Some upper flows that are here temporarily grouped with the Gold Road latite form the general crest of the high ridge east of Oatman and are exposed over much of the slope east of that crest and south of Sitgreaves Pass. They include some layers of light-colored tuff and glassy flows of considerable variety in color and texture. Brown ropy lavas of compact texture are associated with nearly black brittle vitrophyres that are almost as glassy as obsidian and some frothy lavas that are nearly as light and porous as pumice. Some of these upper flows contain rough bubble-like cavities of a peculiar kind known as lithophysae (Greek for stone bubbles). Most of these cavities, some of which are 18 inches long, have been filled with infiltrated silica in the form of chalcedony and opal. When the rock weathers and crumbles the opal and chalcedony in large part remain, scattered in profusion over the ground. These upper members will be excluded from the Gold Road latite and given a distinctive name in the final report.

As seen under the microscope the Gold Road latite is found to be composed generally of fresh clear feldspars, near andesine in composition and showing by absence of zonal structure a fairly uniform composition; biotite; monoclinic pyroxene, pale green or colorless in thin section; with apatite, zircon, magnetite, and, in some specimens, titanite. These crystals lie in a microlitic glassy groundmass. Some specimens show hypersthene, and in a few quartz is present as phenocrysts. Amphibole is very rare. The more glassy varieties show less abundant phenocrysts of the minerals mentioned, in a matrix of volcanic glass that may contain incipient hairlike crystals known as trichites. Many of the glassy rocks display beautifully under the microscope the curved lines of flowage characteristic of some of the more siliceous lavas that have been solidified before complete crystallization could take place.

Chemical analyses of two samples of the Gold Road latite, both made by R. C. Wells in the Survey laboratory, are given below:

Chemical analyses of Gold Road latite.

	1	2		1	2
SiO$_2$	62.96	68.94	H$_2$O below 110° C	0.23	0.28
Al$_2$O$_3$	15.36	13.36	H$_2$O above 110° C	1.37	3.43
Fe$_2$O$_3$	2.57	1.29	TiO$_2$	.72	.47
FeO	2.09	1.04	P$_2$O$_5$	.28	.12
MgO	2.50	.79	MnO	.04	.03
CaO	4.26	2.02			
Na$_2$O	3.84	2.20		100.18	99.86
K$_2$O	3.96	5.89			

1. Three-quarters of a mile southwest of Sunnyside mine, near base of cliff.
2. Three-fifths of a mile south of Sitgreaves Pass, on top of ridge.

The rock represented by analysis 1, although not the conspicuously biotitic variety described as particularly characteristic of the basal flow of the Gold Road latite near Oatman, represents fully as abundant and characteristic a part of the formation. Although this rock may in some places extend down to the base of the formation and in others constitute its upper part, it appears on the whole to occupy a medial position between the biotitic variety and the glassy, ropy, and more siliceous lavas that are generally found near the top of the Gold Road latite as here defined. More specifically, it represents the brownish or reddish-gray, somewhat porous rock that forms the greater part of the cliffs along the western face of Ute Mountain, extending past the heads of Esperanza and Iowa canyons to Sunnyside Wash; it is the rock exposed around the Casey Jones shaft in Antelope Canyon and along the Oatman-Kingman road from Gold Road over Sitgreaves Pass and down to a point within a mile of the pumping station on Meadow Creek; and it is also the principal rock exposed in the lower slopes above Cottonwood Creek up to about a quarter of a mile below Flag Springs. Its general weathering tint is reddish brown. Some of this rock is locally known as rhyolite, and it appears to correspond, in part at least, to Schrader's "rhyolite series."[8] It rarely, however, shows any free quartz, and the analysis is not that of a typical rhyolite. Most of the springs in the district issue from this rock.

Analysis 2 represents certain dark, brittle glassy flows that generally overlie the rock of analysis 1, south of Sitgreaves Pass. The rock is considerably more siliceous than that of analysis 1, with less iron oxide, magnesia, and lime but with more potash and higher total alkalies. It might be called a rhyolite vitrophyre, but, as previously explained, it has not been mapped separately from the other rock in the present preliminary report.

---

[8] Op. cit., p. 40.

The only places in which ore has been found in the Gold Road latite are in the Gold Road and Gold Ore mines, near Gold Road. It is possible that some ore in the Gray Eagle workings of the Tom Reed mine may have been stoped from this rock, but the latite is generally regarded by those familiar with that mine as unfavorable to ore deposition.

YOUNGER VOLCANIC FORMATIONS.

The formations that overlie the Gold Road latite in the northeastern part of the district, namely, the Flag Spring trachyte, the Sitgreaves tuff, the Meadow Creek trachyte, and the Cottonwood rhyolite, contain no ore deposits, so far as known, and will not be here described. Full accounts of them will be given in the final report. Description of the olivine basalt and associated tuffs, the latest products of volcanism in the district, will also be deferred until the complete report is prepared.

MOSS PORPHYRY.

The Moss poryhyry, which is the country rock of the Moss mine, forms an intrusive mass, a little over 2 miles wide, which extends from the vicinity of Silver Creek north-northwestward beyond the limits of the area mapped. The typical rock of the area is a pinkish-gray medium-grained porphyry, in which the poryhyritic texture is not particularly conspicuous, and whose general appearance is suggestive of granite. The recognizable phenocrysts are chiefly plagioclase as much as 1 centimeter in length, and these lie in a crystalline groundmass of pink feldspar (apparently orthoclase), biotite, and amphibole. Under the microscope the larger phenocrysts of sodic plagioclase (oligoclase-andesine) are seen to lie in a fine granular, in part micropegmatitic groundmass of cloudy orthoclase, quartz, biotite, and pale-green amphibole, with magnetite, apatite, titanite, and zircon.

A chemical analysis of typical Moss porphyry from a point by the road to Meals Camp, three-quarters of a mile north of the Moss mine, made by R. C. Wells in the Survey laboratory, is as follows:

Analysis of Moss porphyry.

SiO_2	62.54	H_2O under 110° C	0.29
Al_2O_3	14.42	H_2O above 110° C	.84
Fe_2O_3	3.51	TiO_2	1.05
FeO	2.57	P_2O_5	.36
MgO	2.41	MnO	.07
CaO	4.26		
Na_2O	3.83		100.13
K_2O	3.98		

The mineral and chemical compositions are those of a quartz monzonite porphyry. The chemical analysis is nearly identical with that of the Gold Road latite (No. 1, p. 26). The Moss porphyry is younger than the Alcyone trachyte and, although definite evidence on this point was not obtained, is believed to be younger than the Oatman andesite. It is possible that the rock, as suggested by its chemical composition, may be the intrusive equivalent of the Gold Road latite.

A number of variations from the typical rock are to be noted. Where the porphyry is intrusive into the Alcyone trachyte, as west of the Moss mine, the contact is in many places very inconspicuous, as the porphyry grades marginally into varieties that closely resemble the trachyte. A quarter of a mile south of the Mossback mine, in the normal Moss porphyry, are some irregular masses of a more coarsely crystalline rock with large pink orthoclase crystals. This rock, which is probably a local variation of the Moss porphyry, has not been separately mapped on Plate I.

The area mapped as Moss porphyry south of the north branch of Silver Creek is occupied, as previously noted, by a much bleached and altered rock that is only in small part identifiable. It probably includes some Gold Road latite and considerable porphyry that is different in character from that near the Moss mine but is considered as probably belonging to the same mass. Much of this material along Silver Creek, near the old stone cabins of the early settlers, is a coarse breccia that weathers to a rather bright coppery green. The origin of this breccia is not clear, but the most reasonable explanation found is that the Moss porphyry, probably carrying in this vicinity some inclusions of the volcanic rocks, was shattered by the continued movement of the intrusive mass after it had in part solidified. Breccia and massive porphyry appear to be mingled in inextricable confusion.

TIMES PORPHYRY.

The exposures of Times porphyry are confined to the triangular area between Silver Creek and Times Gulch, where the rock forms a very irregular body intrusive into the Alcyone trachyte, into which it has sent many little dikes, and, on the north, into the Moss porphyry. Hardy Mountain, Cathedral Rock, and the bold pinnacles along the north side of Times Gulch (Pl. IV, A) west of the Times mine are all composed of this rock, which in general is easily recognized from a distance by its pale-pinkish tint and massive weathering.

The Times porphyry is a faintly pinkish rock of rather fine texture and not conspicuously porphyritic. At first glance it appears to be composed almost wholly of feldspar with ill-defined phenocrysts, the largest 1 centimeter in length, of the same mineral and a few small ragged flakes of biotite. A hand lens shows, however, that quartz is abundant in the groundmass, although absent as distinct crystals.

Under the microscope, in thin section, the Times porphyry is seen to be composed mainly of quartz and orthoclase. These minerals do not as a rule occur in distinct separate crystals or grains but are intimately intergrown in the manner usually described as micropegmatitic or micrographic. Such intergrowths as seen through the microscope in polarized light with nicols crossed are striking objects, and the texture is rarely more beautifully exemplified than in the Times porphyry. A few thin sections show considerable areas of quartz and orthoclase that are not intergrown, but these areas rarely show any crystal boundaries and generally grade irregularly at their margins into the characteristic micrographic intergrowth. The mineral composition of the Times porphyry is very simple. In addition to the quartz and orthoclase, most sections show a little sodic plagioclase, biotite, apatite, and magnetite.

A chemical analysis of the Times porphyry, on material collected on the north side of Times Gulch, 1¾ miles south of Cathedral Rock, has been made by R. C. Wells in the Survey laboratory. The results are as follows:

Chemical analysis of Times porphyry.

SiO_2	72.85	H_2O under 110° C	0.08
Al_2O_3	13.45	H_2O above 110°C	.40
Fe_2O_3	1.11	TiO_2	.39
FeO	.54	P_2O_5	.04
MgO	.24	MnO	.06
CaO	1.13	CO_2	.27
Na_2O	4.01		
K_2O	5.40		99.77

The analysis, although the proportion of soda is a little high, corresponds to that of granite, and the rock is classed as granite porphyry. The visible plagioclase seems insufficient to account for all the soda, and presumably the orthoclase is sodic. Whether it is soda orthoclase or a microperthitic intergrowth of normal orthoclase and albite could not be definitely determined, as the orthoclase is generally too turbid for satisfactory optical examination.

The analysis of the porphyry is nearly identical with that of the glassy spherulitic flow which in this report is named the Cottonwood rhyolite, and probably the rock is a product of the same magma. As the rhyolite, next to the basalt, is the youngest of the flows, this would indicate that the Times porphyry is also one of the younger rocks of the district.

The Times porphyry could be quarried in blocks of sufficient size for building stone were there any local demand for such material, and it should be readily worked and durable.

DIKES AND SMALL INTRUSIVE BODIES.

Dikes of various kinds are abundant in the Oatman district and may be classified roughly, in advance of detailed study, into felsite porphyry [9] and basalt. Under felsite porphyry are included numerous light-colored dike rocks, most of which show phenocrysts of orthoclase and quartz and are generally termed rhyolite in the district. A few are darker, evidently less siliceous rocks and may be andesite porphyry or latite porphyry. Narrow buff or pale greenish-gray dikes of various trends and many of them sharply curved and branching traverse the Moss porphyry. As a rule, the rock of these dikes is not particularly fresh, and determination of its exact lithologic character is perhaps not possible.

One rather large dike, in places 50 feet wide, can be traced from the western boundary of the district, half a mile south of Silver Creek, across Silver Creek at the old stone cabins and thence in a general east-northeasterly direction to a point two-thirds of a mile south-southeast of the Mossback mine. This is generally a light-gray or greenish-gray rock with conspicuous phenocrysts of quartz and orthoclase. In surface exposures it is as a rule decomposed, and crumbling well-formed Carlsbad twins of orthoclase can be picked from the soft disintegrated dike rock.

A similar but generally harder and fresher rock forms a dike that can be followed from the western boundary of the district $1\frac{1}{4}$ miles north of Times Gulch eastward past the Times and Ivanhoe mines to a point about a quarter of a mile north of the Blue Bird mine. West of the Times mine this dike follows the contact between the Times porphyry on the south and the Alcyone trachyte on the north. It is conspicuously porphyritic, and rounded phenocrysts are particularly abundant in it.

Many smaller, more irregular, and less persistent felsite dikes occur in the central part of the district. Most of them trend from west to west-northwest, and some of them split into many branches. In tracing one of these dikes it is rather common to find that the individual dike followed comes to an end but that another can be picked up a short distance on either side of the one first followed. This second dike can be followed for some distance to another offset, and so on. A typical example of one of these small felsite dikes is crossed by the Oatman-Kingman road between the United Western and Arizona Central shafts and can be followed westward past the

---

[9] In accordance with L. V. Pirsson's suggestion (Rocks and rock minerals, p. 202, New York, 1908) for the classification of rocks without the aid of the microscope.

Blue Bird, North Star, and Midnight mines to a point about a quarter of a mile north-northwest of the Oatman Belle shaft and perhaps beyond.

The basalt dikes occur mainly in the northeastern part of the district and as a rule are narrow but remarkably regular and persistent. Their general trend is from north to north-northwest. Some of them plainly cut some of the basalt flows, but it is probable that these flows themselves were fed through similar fissures and that the dikes which now traverse them were the feeders of higher flows that have been removed by erosion.

Small bunchy, irregular intrusive bodies of a nearly white rhyolite are rather abundant in certain parts of the Oatman district, particularly in the Gold Road latite east of Oatman, and have produced two of the most conspicuous topographic features of the region. One of these is the sharp white spine locally known as The Elephant's Tooth, or The White Elephant, which is visible in eastward views from nearly all points near the town of Oatman (Pl. II, B); the other is the still more imposing Boundary Cone (Pl. V, B). These peaks are plugs which were forced up in molten condition through the surrounding rocks and which, owing to the resistance of their material, have been left as towers after the once inclosing rock has been stripped away by erosion. The Elephant's Tooth plug, as exposed at the surface, is oval in plan and only about 600 feet long. Northwest of it are a number of other intrusive rhyolite masses, the largest of which, about a quarter of a mile northeast of Oatman, is about 2,500 feet long and 800 feet wide. Most of these masses are nearly white glassy rhyolite, showing flow structure. Closely associated with these rhyolite bodies are some small masses of rather peculiar breccia. Some of this breccia in addition to rhyolitic fragments contains angular pieces of the Gold Road latite. The occurrence of the breccia in relatively small masses within the latite suggests that it is, in part at least, an intrusion breccia. It may be supposed that some of the small rhyolitic masses were intruded under a rather thin cover of older rock and, being stiff and viscous, were partly brecciated by the movement of the lava. Probably also there were volcanic explosions which shattered the inclosing latite.

Southeast of Oatman, in the vicinity of the Kokomo shaft of the Oatman Mines Co., is a rock intermediate in character between the Gold Road latite and typical rhyolite. It is intrusive into the latite but is closely associated with breccia and with glassy lithophysal lavas that can not be satisfactorily separated from some of the flows of similar material that make up much of the upper part of the formation mapped as Gold Road latite. The rhyolitic rock near the

Kokomo shaft has not been separated from the Gold Road latite on the preliminary geologic map (Pl. I).

The Boundary Cone plug (Pl. V, *B*) is a fine example of this form of intrusive body. The tower rises about 1,500 feet above the general level of the country west of it. A prominent dike of rhyolite extends westward from the central mass nearly to the Oatman-Topock road,[10] where it is covered by the Quaternary alluvial deposits. A nearly horizontal offshoot, or sill, also projects into the Alcyone trachyte on the south side of the cone. A smaller plug, surrounded by the Alcyone trachyte, occurs just northeast of the main mass.

THE GOLD-BEARING VEINS.

DISTRIBUTION, STRIKE, AND DIP.

There is in the Oatman district no general distinction to be made between faults and veins. Practically all the veins occupy fissures along which faulting has taken place, as a rule before, during, and after the period of vein formation. On the other hand, some fissures, such as the Mallery fault, are younger than the veins and contain no original ore.

Vein-filled fissures are particularly abundant in the southern and central parts of the district, especially in the Oatman andesite and Esperanza trachyte. The general strike is northwest, but the strike of individual fissures may range from nearly north to nearly west. The dip as a rule is high, over 60°, and to the northeast. Many of the fissures branch, but there appears to be no recognizable rule in respect to this; some fissures diverge to the northwest, as near Old Trails, and others to the southeast, as on Leland Hill. The veins vary in width, but few of them are wider at any point than 50 feet. Some are of simple tabular form, with well-defined walls; others are complex lodes, the main mass of vein material being accompanied by many branching, irregular stringers; others are aggregations of stringers, or stringer lodes. Some crop out boldly and conspicuously; others are most insignificant as surface features. As a rule the large ore bodies of the district have been mined from veins or parts of veins that are not prominent at the surface.

MATERIAL AND STRUCTURE OF THE VEINS.

The Oatman veins are mineralogically of simple character, consisting mainly of quartz, calcite, and adularia, associated, in the ore shoots, with free gold. As a rule only quartz and calcite are recognizable with the naked eye. The adularia occurs generally in

---

[10] The old road. The new road crosses the dike.

A. TWO MILES SOUTH-SOUTHWEST OF OATMAN, LOOKING NORTHWEST.

B. NEAR THE TREADWELL SHAFT OF THE PIONEER MINE.

LAMINATED STRUCTURE OF THE ALCYONE TRACHYTE.

A. NEAR VIEW OF THE GOLD DUST VEIN IN A SURFACE CUT JUST NORTH OF THE NO. 2 SHAFT OF THE GOLD DUST MINE.

The principal mass of quartz and calcite is here about 5 feet wide, with many stringers on both sides.

B. MORE GENERAL VIEW OF THE GOLD DUST VEIN AT THE SAME LOCALITY, LOOKING NORTHWEST.

The extraordinary complex of quartz-calcite stringers here displayed is on the hanging-wall side of the vein shown in *A* and forms a zone at least 100 feet wide. Calcite is more abundant than quartz.

VEIN STRUCTURE.

A. PART OF THE PIONEER VEIN AT THE TREADWELL SHAFT, LOOKING NORTH.

Shows a complex of banded quartz-calcite stringers.

B. PIONEER VEIN, LOOKING SOUTH ALONG THE VEIN AND OVER THE TOP
OF AN OLD STOPE NEAR THE THIRTY-FIFTH PARALLEL SHAFT.

This vein shows the smooth, regular hanging wall of a strike fault that is younger than the vein
and supposedly cuts off the ore bodies above the 400 level.

VEIN STRUCTURE.

A. THE GOLD ROAD VEIN AS EXPOSED ON THE OATMAN-KINGMAN ROAD, LOOKING SOUTHEAST.

Shows the relatively regular zone of sheeting followed by this vein.

B. THE MOSS VEIN AT THE MOSS MINE.

One of the most massive quartz outcrops in the district.   Fluorite appears to be generally more abundant than calcite in this vein.

microscopic crystals, and gold is visible only in unusually rich ore. Sulphides, even pyrite, that almost invariable constituent of the ores of other districts, are absent from the typical Oatman ores, although a little chalcopyrite and chalcocite have been found in the Gold Ore vein, near Gold Road, and Schrader [11] reports that pyrite was present in the small ore body of the Rattan mine. The country rock near the veins may contain finely disseminated pyrite in small crystals, but it is nowhere conspicuous or abundant. Fluorite occurs in some of the veins, particularly those in the Moss and Times porphyries, but apparently is not particularly significant as to the presence or absence of gold. In the rich ore from the upper workings of the Moss mine some of the gold was found embedded in fluorite, and gold was associated with fluorite in a narrow but high-grade ore shoot stoped in the Buckeye vein, a spur from the Hardy vein. On the other hand, fluorite, although a little was seen in 1921 in a small isolated ore shoot in the Aztec mine, is very rare in the larger ore bodies of the district. In the northern part of the district the mineral is found abundantly in some veins, which so far as known are practically barren.

The proportion of quartz and calcite in the veins varies widely. Some veins are chiefly quartz; others chiefly calcite. A wide range also may be found in different parts of the same vein. As a rule the ore is found where both minerals are present. Veins or parts of veins that consist entirely of quartz or calcite are, in the Oatman district, generally of very low grade or barren.

The larger veins in the Oatman district are essentially stringer lodes of very complex structure. Some idea of their complexity may be had from figure 2, drawn from sketches, made in March, 1921, of the Big Jim vein as exposed in crosscuts on the 400 level. It will be noted that what is generally termed "the vein" is actually made up of two or more veins (the term being here used with its narrower meaning for the filling of a fissure), together with many irregular veinlets or stringers and a considerable proportion of more or less altered andesitic country rock.

The appearance of some typical veins in surface exposures is shown in Plates VII–IX.

The individual veins and stringers appear banded in cross section, showing that the vein minerals were deposited in successive layers from the walls to the middle of the fissure. Such banding was particularly noticeable in the large ore body of the United Eastern mine from the 700 level up, and a typical specimen of the ore from this

---

[11] Op cit., p. 172.

vein is shown in figure 3. In many of the larger stringers or individual veins of the lode the banding is interrupted, and it is plain that earlier vein material has been shattered and then cemented by later vein material. In the United Eastern and Big Jim veins the shattered vein material consists in large part of gold-bearing, banded quartz and calcite, and the later material of relatively barren calcite.

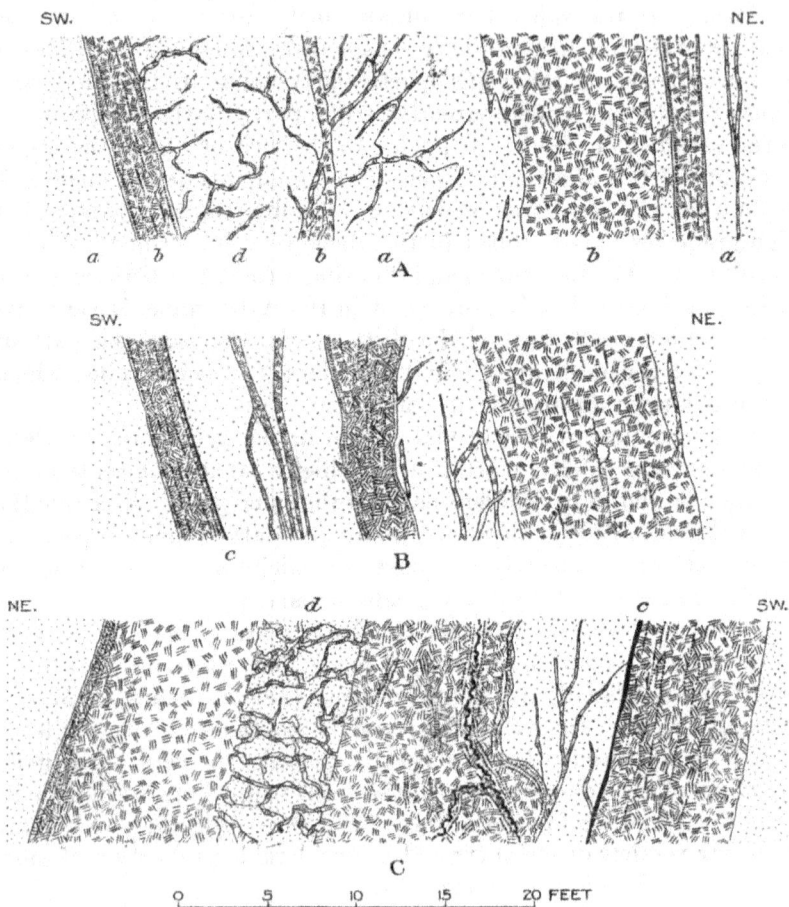

FIGURE 2.—Sketches showing the structure of the Big Jim vein on the 400 level. *a*, Oatman andesite; *b*, quartz and calcite, in places showing a banded structure; *c*, gouge; *d*, andesite traversed by many irregular stringers of quartz and calcite.

In a broad way in the Oatman veins the deposition of fine-grained white quartz, which has, in part at least, replaced andesite and contains little or no gold, has been followed by the deposition of the gold-bearing quartz accompanied by some calcite and adularia, followed in turn by barren calcite. This general sequence, however, is certainly far from being the complete record. Much of the quartz that was deposited nearly or quite contemporaneously with the gold has clearly replaced older calcite. Some of it, moreover, appears to have crystallized simultaneously with calcite. This indicates at least three generations of calcite. A specimen collected from the Big Jim dump shows also three distinct generations of quartz, all of which are probably older than the characteristic gold-bearing quartz. The

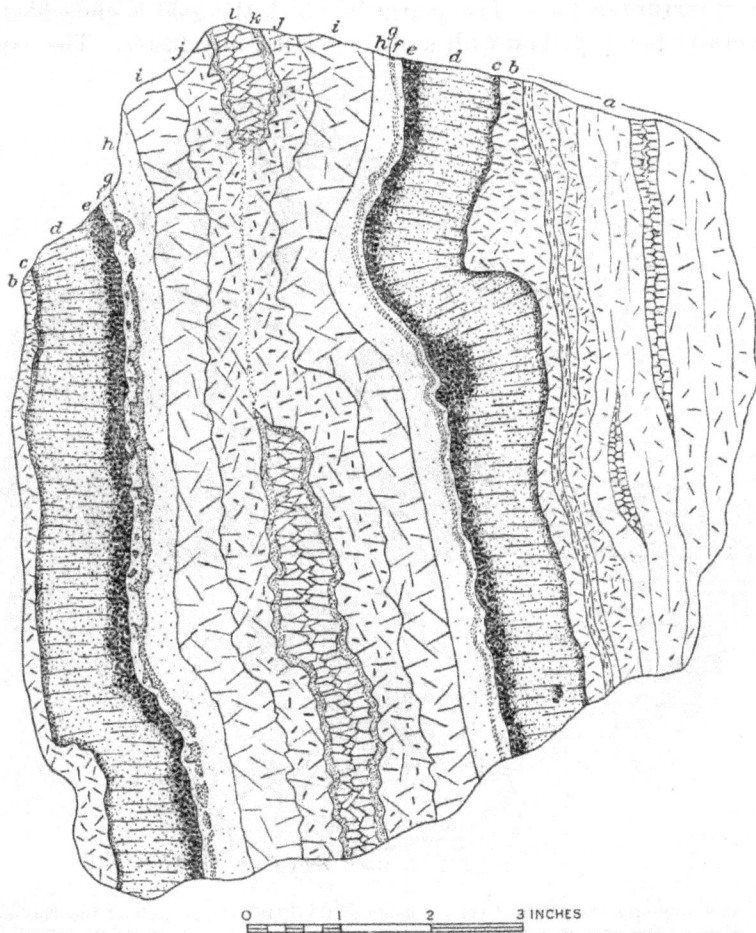

FIGURE 3.—Typical banded ore from the stope above the 600 level of the United Eastern mine. a, Brown quartz in thin crusts; b, rather finely crystalline white milky quartz; c, thin layer of honey-yellow oily-looking quartz and probably also microscopic adularia; d, clear quartz intimately associated with the honey-yellow variety and with calcite, probably also microscopic adularia; e, layer of dark calcite, quartz, and probably adularia; f, pale greenish-white fine-grained quartz; g, thin layer of honey-yellow, oily-looking quartz; h, like f; i, rather coarsely crystalline milk-white quartz; k, oily-yellow quartz; l, clear comb quartz.

conclusion reached is that during the middle stage of vein formation quartz and calcite were repeatedly deposited alternately and that during this period also they were at times deposited simultaneously and some calcite was replaced by quartz. Deposition of calcite has probably continued up to the present time.

The quartz of the Oatman veins is generally fine grained, and some of it, particularly in the finely banded Gold Road vein, is chalcedonic or flinty in texture. Crustification is common and gives the veins and stringers a banded appearance as seen in cross section, due to the fact that successive crusts or layers of quartz differ in texture or hue. The quartz in which the gold is embedded is generally honey-yellow with a characteristic oily luster. The cause

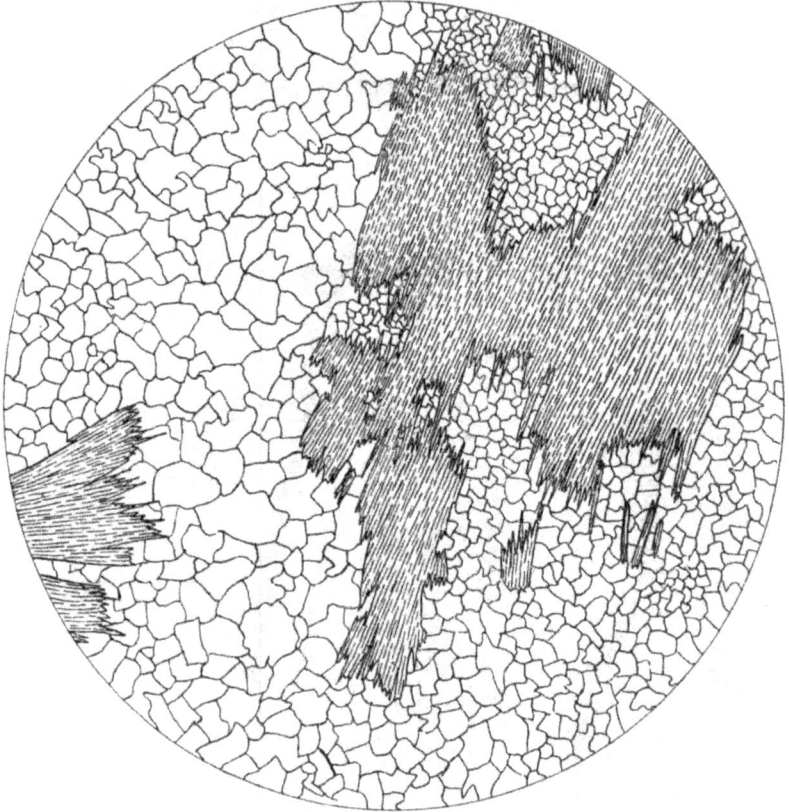

FIGURE 4.—Quartz and calcite from the dump of the Black Eagle shaft of the Tom Reed workings. Drawn with camera lucida from a thin section as seen under the microscope. The calcite (lined shading) in ragged areas is believed to be residual, an originally calcitic vein having been partly replaced by granular quartz. Magnified 15 times.

of this tint and luster has not yet been ascertained, as the microscope throws no definite light on this peculiarity.

Much of the quartz associated with the gold in the Oatman veins has a lamellar structure. In some varieties the quartz is compact and fine grained, but when seen in thin section under the microscope with the polarizing nicols crossed the structure becomes strikingly apparent. Each lamella or leaf of quartz consists of granules which terminate sharply against rectilinear boundaries that are invisible in ordinary light. The structure is such as might be expected had the quartz crystallized between closely spaced parallel or slightly divergent impervious plane partitions so thin as to be invisible. In other varieties the quartz forms fragile septa with open spaces between, so that the mass as a whole is cellular or porous. These structures have been recognized as characteristic of the replacement of calcite by quartz, and there is ample evidence that such is their

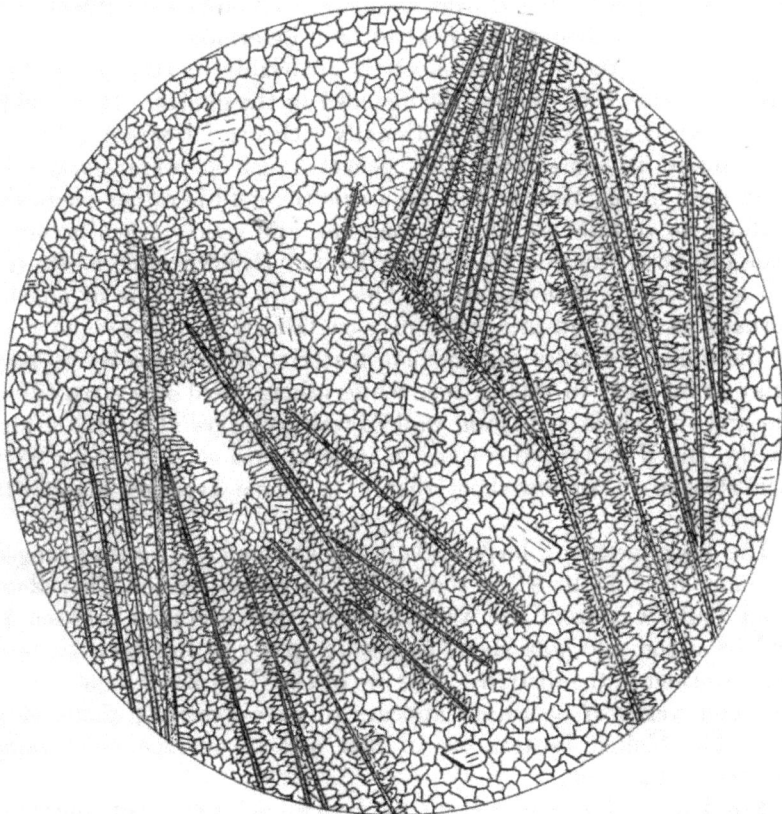

FIGURE 5.—Lamellar aggregate of quartz, calcite, and adularia from the dump of the Black Eagle shaft of the Tom Reed workings. Drawn with camera lucida from a thin section as seen under the microscope. The remarkably thin plates of calcite are fringed with bristling quartz crystals and separated by granular quartz (clear) with some adularia (light-line shading). The section shows one small vug or cavity lined with quartz crystals. Magnified 15 times.

origin in the Oatman district. The relation of quartz and calcite in the Oatman veins, however, presents some problems that will be discussed in the complete report. Figures 4 and 5 illustrate two occurrences of quartz and calcite in the Tom Reed vein.

Adularia, a clear variety of the potash feldspar orthoclase, is commonly present with the gold-bearing quartz but can rarely be recognized with the naked eye. Under the microscope, in thin sections, what at first looks like a granular aggregate of clear quartz grains is seen on close inspection to contain grains with a lower index of refraction and lower birefringence than quartz. Some of these grains are of irregular form, but many show more or less of the characteristic rhomboidal or lozenge-shaped outline of adularia and fine cleavage lines. (See fig. 5.)

The calcite of the veins is generally white and shows no unusual features. Whether it contains manganese or other constituents besides calcium carbonate has not yet been determined.

The gold of the large ore shoots is invariably in very small clustered particles, like yellow dust included in the quartz. It is visible to the naked eye only in exceptionally rich ore. Coarser gold has been found in some small rich pockets, such as were worked in early days in the Moss mine. As a rule the gold is rather narrowly confined to certain portions of the vein, and assays taken across a wide ore body, such as that of the United Eastern mine above the 700 level, show great variation. One or two thin layers of quartz, perhaps less than an inch wide, may contain by far the greater part of the gold. Samples that include material from these layers may assay from $100 to $200 a ton, while others yield less than $10 a ton and the whole vein from wall to wall averages about $25 a ton. In the Big Jim vein there are in general two such comparatively rich streaks in an ore shoot that as a whole averages about $16 a ton.

The ore mined on a large scale in the Oatman district has ranged in tenor from $7 to $35 a ton. The ore mined by the Tom Reed Gold Mines Co. from the Aztec mine in 1921 averaged between $7 and $8 a ton and was about as low in grade as has been successfully worked in the district. The ore of the Gold Road mine during the later years of operation appears to have averaged about $9 a ton. The United Eastern ore in 1921 had an average assay value of nearly $22 a ton.

The United Eastern ore, up to the end of 1921, has contained from one-half to two-thirds as much silver as gold, by weight. So far as known the silver occurs only as the usual natural alloy with gold.

ALTERATION OF THE WALL ROCK.

The changes effected in the wall rock by the solutions that deposited the veins have not yet been carefully studied and can not here be fully described. In general, the Oatman andesite near the veins is light greenish gray and is softer than the original fresh rock. The feldspar phenocrysts are dull and chalky, and the groundmass contains disseminated crystals of pyrite, generally 1 millimeter or less in diameter. Thin sections of such rock, seen under the microscope, show that the biotite has become pale green or has been altered to aggregates of chlorite; the feldspars are now aggregates of calcite, quartz, and sericite. Augite has been replaced by calcite, and calcite and quartz are abundant throughout the rock in irregular aggregates. The iron originally present in the rock in biotite, augite, and magnetite is now combined in chlorite and pyrite. The only original mineral remaining is apatite.

NOTES ON SOME OF THE MÓRE IMPORTANT VEINS.

TOM REED VEIN.

The Tom Reed vein, which, broadly considered, includes the American, Aztec, Big Jim, original Tom Reed or Benjamin Harrison, and United Eastern veins, has been by far the most productive in the district, having yielded to the end of 1921 gold and silver to the value of nearly $20,000,000.

The general course of this vein and its relation to the geologic formations of the district are shown in Plate I. The principal mine workings upon it appear in Plate X, and a view over the surface, along the lode, is reproduced as Plate XI, *A*. Another view, in the opposite direction, is shown as Plate XI, *B*.

The general course of the vein, like that of most others in the district, is approximately northwest. On the southeast it appears as a zone of fissuring and faulting that crosses the crest of the main mountain ridge about 2 miles southeast of Oatman and, apparently as two nearly parallel fissures, about 250 feet apart, continues down the steep west slope of the ridge past the Argo mine. Thence it is generally though not continuously traceable on the surface past the shaft of the United American mine and can be similarly followed northwestward nearly to the Aztec shaft of the Tom Reed Gold Mines Co. From the Aztec mine the vein is continuous underground through the Big Jim workings (see Pl. X), but its relations at the surface are complicated by a dislocation along the Mallery fault. The vein in the Aztec and Big Jim workings dips 70°–75° NE., whereas the Mallery fault, striking nearly parallel with the vein, dips 60°–70° SW. The fault is normal, with a throw of about 400 feet. The effect of the fault is to throw the upper part of the Tom Reed vein down and to the southwest, as illustrated by the cross sections of figure 6. As shown by these sections, the part of the vein under the fault plane has no outcrop but is limited above by the fault plane, from 100 to 400 feet below the present surface. On the other hand, the part of the vein above the fault plane does theoretically come to the surface, although in places it is obscure or covered by superficial detritus. It is this portion above the fault plane that was originally developed and stoped by the Tom Reed Gold Mines Co. through its Grey Eagle and Bald Eagle shafts. (See Pl. X.) The discovery that the Grey Eagle and Bald Eagle ore

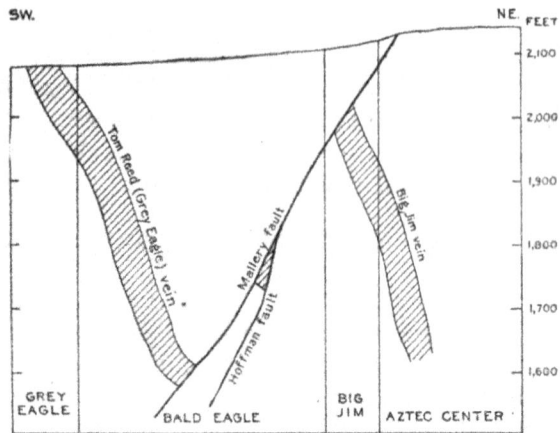

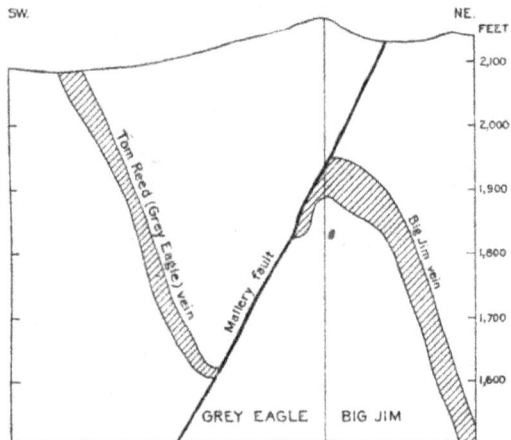

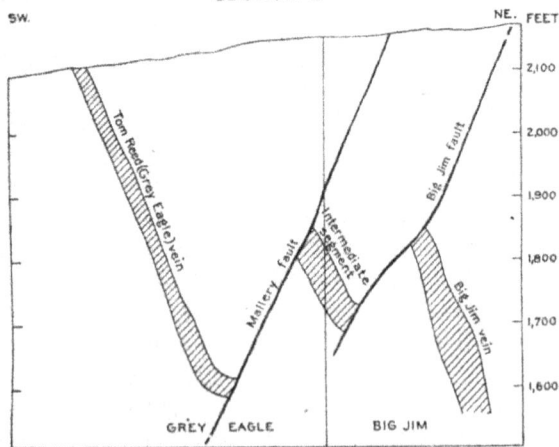

FIGURE 6.—Cross sections of the Tom Reed-Big Jim vein and Mallery fault. Taken, with omission of some details, from a series of sections prepared by O. H. Hershey, consulting geologist, and Herman Zulch, engineer, for the Tom Reed Gold Mines Co. Section 3 is about 400 feet southwest of the Grey Eagle shaft, section 19 is through that shaft, and section 29 is about 250 feet northwest of the Grey Eagle shaft and about 80 feet southeast of the Big Jim shaft.

A. VIEW NORTHWEST, TOWARD OATMAN, ALONG THE TOM REED VEIN FROM
THE VICINITY OF THE AZTEC SHAFT.

1, Aztec shaft; 2, Bald Eagle shaft; 3, Grey Eagle shaft; 4, Red Lion shaft; 5, Big Jim shaft;
6, Benjamin Harrison shaft; 7, Olla Oatman shaft; 8, United Eastern No. 1 shaft; 9, United
Eastern No. 2 shaft.

B. VIEW SOUTHEAST ALONG THE TOM REED VEIN.

1, Grey Eagle shaft; 2, Hooper shaft, on outcrop of the upper segment of the Tom Reed vein;
3, Bald Eagle shaft, on same outcrop; 4, Aztec shaft; 5, Black Eagle dump; 6, American
mine.

MINES OF THE OATMAN DISTRICT.

A. VIEW UP GOLD ROAD GULCH, SHOWING THE SURFACE RELATIONS OF THE GOLD ROAD MINE.

1, Gold Road mill; 2, No. 1 shaft; 3, No. 3 shaft. All the rock included in the view is the Gold Road latite. The general easterly dip of the flows is distinctly shown.

B. LELAND HILL FROM THE SOUTHEAST, LOOKING ALONG THE VEIN.

1, No. 1 tunnel; 2, No. 2 tunnel; 3, No. 3 tunnel; 4, Mitchell shaft.

MINES OF THE OATMAN DISTRICT.

was cut off below by the fault and the opening up of the deeper part of the vein in the Big Jim ground led to the litigation between the Tom Reed Gold Mines Co., owner of the Bald Eagle and Grey Eagle claims, and the United Eastern Mining Co., owner of the Big Jim claim. There can be no real doubt that the Grey Eagle vein was once the direct upward extension of the Big Jim vein. The question at issue was whether such former continuity confers a legal right on the owner of the older Grey Eagle and Bald Eagle claims to exploit the part of the vein under the fault plane, within the Big Jim claim.

Within the boundaries of the Big Jim, Grey Eagle, and Bald Eagle claims the Mallery fault has been carefully studied by geologists engaged by the mining companies, particularly by Oscar H. Hershey, for the Tom Reed Co. Its position has been mapped, and its effects are generally well understood. It is accompanied by other generally subordinate faults, some of which, however, in certain cross sections nearly equal it in displacement. The effects of the Mallery fault and its subsidiary dislocations are illustrated generally in figure 6, which has been reduced and simplified from the series of elaborate cross sections prepared by Mr. Hershey in connection with the lawsuit. The Big Jim fault, probably a branch from the Mallery, is found in sections north of the Grey Eagle shaft. Between it and the Mallery fault, as shown in section 29 of figure 6, is a portion of the vein which has been called by one party to the litigation the "sideline vein," and by the other the "intermediate segment." Considerable ore has been stoped from this segment, and as it lies generally under the side line between the Grey Eagle and Big Jim claims, ownership was claimed by both parties to the suit. The decision rendered in the Superior Court of Arizona at Kingman, on March 28, 1921, asserted the existence of three distinct veins—the Tom Reed vein, the sideline vein, and the Big Jim vein—although it was acknowledged that "at some time in the dim and distant past" they had been continuous. It gave the Tom Reed vein to the Tom Reed Gold Mines Co. and the Big Jim vein to the United Eastern Mining Co. That part of the "sideline vein" that has its apex wholly within the ground of the Tom Reed Co. was decided to belong to that company; all other parts to the United Eastern Co. The case has been appealed and was argued before the Supreme Court of Arizona at Phoenix in April, 1922. Decision had not been rendered at the date of writing.

South of the Grey Eagle shaft another subsidiary fault, known as the Hoffman fault, lies under the Mallery. Between the two fissures some small bodies of ore have been found. A number of minor faults associated with the Mallery are known from underground work but will not be described in this preliminary report.

Where exposed in the mines the Mallery fault is generally easily recognizable. As a rule it shows a seam of reddish clayey gouge, rarely over an inch thick, lying between smooth slip surfaces and accompanied by about a foot of soft, crushed andesite or latite. In places there are two or three seams of red gouge, and the neighboring rock may be fissured and decomposed for a width of 2 or 3 feet. There has been no deposition of vein material in the fault fissure, although near the places where it cuts the Tom Reed vein it generally contains some dragged ore. As a whole the fault itself is a rather inconspicuous feature in view of its considerable throw of about 400 feet. Whether the displacement is a simple dip slip or involves also a considerable strike slip—that is, has a notable horizontal component parallel with the strike—is not known. The lack of definite information on this point was brought out in the hearings before the superior court in Kingman, and the judge held that it is by no means certain that the part of the vein in the Grey Eagle claim was ever a direct upward continuation of the part now within the Big Jim claim.

At the surface the Mallery fault is very obscure. It has been exposed definitely in a small shaft under the aerial tramway to the Aztec mine, about 230 feet east of the Grey Eagle shaft. Soft, decomposed zones that probably mark the outcrop of the fault have been opened also by a few shallow cuts and pits on the Big Jim and Aztec Center claims. It probably lies a short distance west of the Hartman tunnel, a disused opening just south of the Aztec shaft, and passes under a little saddle between the tunnel and the Black Eagle shaft. The position of the Mallery fault near the Black Eagle shaft is not definitely known. A trip down this shaft, which is no longer used, failed to reveal the fault. The saddle which lies just southwest of the American shaft and through which passes the road to the Argo shaft is certainly determined by erosion along a fault that has dropped the Gold Road latite on the west against the Oatman andesite on the east. This may be the Mallery fault. It is traceable, though not continuously, in a general south-southeasterly direction through the Lodestone and Sunny Side claims and apparently passes just west of the Sunny Side shaft. Thence it turns to the southeast and can be followed with some measure of probability for about 1½ miles.

North of the Big Jim shaft the position of the Mallery fault is also in doubt. It is apparently cut and displaced by a generally west-northwest fissure, which is shown on Plate I as the Oatman fault. This fault apparently has a throw of some hundreds of feet, with the downthrow on its north side. To the east it can be traced over the main ridge and down the north side of Antelope Canyon.

To the west it presumably meets the Tom Reed vein obliquely and may be represented by the post-ore gouge seams along the Tom Reed and United Eastern veins. The ground where the Oatman fault and the Tom Reed vein supposedly meet, mainly under the Rising Star claim, is covered by superficial detritus and for the most part unexplored by mining operations. The present accessible underground workings throw no light on the relation between the Mallery and Oatman faults.

What becomes of the Mallery fault north of the Big Jim workings? This question has not been finally disposed of. Some have been inclined to believe that the Mallery fault becomes coincident with the Oatman fault and is represented by the gouge-filled fissures that accompany the Tom Reed and United Eastern veins. This, however, is not probable, for the dip and throw of the Mallery fault are to the southwest, whereas the dip and throw of the Oatman fault are to the northeast. It is more likely that the Mallery fault is represented by a fissure, shown on Plate I, which has been opened by some cuts and tunnels north of the Big Jim shaft and is traceable over the surface on the North Aztec, Emma, and Merrill claims, past the Oatman United shaft. This fissure apparently was the incentive for sinking the Oatman United shaft at the spot chosen and was followed by a drift for about 600 feet on the 600 level. As exposed on that level, the fissure contains a red gouge and shows a rather close resemblance to the Mallery fault as exposed in the Aztec and Big Jim workings. It contains a little calcite, but, so far as known, no ore. The general dip appeared to be between 70° and 75°, to the southwest. North of the Oatman United shaft the fissure, although in places obscure on the surface, appears to continue across the Tonopah No. 1 claim into the mass of intrusive rhyolite that occupies portions of the Yankee, Dollar Princess, Surprise, and other claims east of the United Eastern mine. It could not be traced north of the rhyolite, but the displacement appears to be taken up by another fissure that forms the eastern boundary of the rhyolite and continues north-northeastward, through the United Northern group of claims. Here, about half a mile southwest of Gold Road, this fissure dips southwest and drops the Gold Road latite on the southwest against the Oatman andesite on the northeast. The throw is roughly determinable as about 400 feet. This fault thus corresponds in dip and throw to the Mallery fault. The probability that it is identical or, at least, that it is a member of the same fault zone is increased by the fact that most of the fissures in this part of the district dip northeast and the downthrow is generally to the northeast, as along the Gold Road and United Eastern veins.

Under the town of Oatman the Tom Reed vein branches. The southwest branch, which to all appearances is the main vein, shows along most of its course a prominent outcrop of quartz and calcite. It represents a zone of fissuring along which the fissures have been filled with quartz, accompanied by some silicification of the Oatman andesite. Subsequently this material was shattered and cemented with calcite, which also forms separate veinlets. In places north of the Olla Oatman shaft this outcrop is from 30 to 40 feet wide and consists chiefly of calcite. Where the vein is crossed by the south branch of Rattlesnake Gulch it shows a change of dip. South of this point the dip is northeast, but in Rattlesnake Gulch it is southwest, at 80°. The width in the gulch is from 6 to 8 feet. In the north branch of Rattlesnake Gulch the lode consists of a vein of calcite only 3 feet wide, which dips 75° SW. For at least 100 feet in the hanging wall the andesite is traversed by many stringers of calcite. From this locality northwestward the vein becomes less conspicuous and within a short distance ceases to be recognizable as a surface feature. This southwest branch of the Tom Reed vein has been explored from the Olla Oatman and Ben Harrison shafts on the 300 and 500 levels for a distance of more than 2,200 feet, but no ore in economic quantity has been found.

The northeast branch of the Tom Reed vein, or the United Eastern vein, in contrast with the other branch, is marked by no outcrop of vein material and probably would not have been recognized as a vein had it not been discovered by mining operations.

In April, 1921, it was still possible to reach the Tom Reed 500 level through the fast-closing, abandoned stopes of the United Eastern mine and to see the place where the two veins separate. There was no ore at this point, which is about 200 feet northwest of the most northwesterly stope of the Tom Reed (Ben Harrison) mine. The main vein, consisting here chiefly of many irregular stringers of calcite accompanied by a strong gouge-filled fissure, plainly divides into two branches, each similar to the main vein and each accompanied by a gouge-filled fissure. Nothing was seen to indicate that one branch is older than the other. The angle of divergence is about 50°, and as this angle is nearly bisected by the strike of the vein south of the point of branching, one branch is as much to be regarded as the direct continuation of the main vein as the other. As already mentioned, the Tom Reed Co. followed what was afterward known as the United Eastern vein for about 100 feet, carrying the drift a short distance beyond the Tom Reed ground. Had the operators continued a few feet farther they would have discovered the largest and richest ore body yet found in the district, but they evidently did not consider the calcitic stringer lode sufficiently promising to justify purchase of the Tom Reed Extension claim, now owned by the United Eastern Mining Co., or to warrant further exploration of ground beyond their own boundaries. Subsequent exploration was accordingly confined to the southwest branch, covered by the Thomas B. Reed, Olla Oatman, and Bessel claims, which so far as now known is barren.

The United Eastern vein nowhere appears at the surface as a distinct vein, although for about 1,800 feet northwest of the point of departure from the Tom Reed vein the fissure zone which the United Eastern vein occupies coincides, as nearly as can be determined from the unsatisfactory surface exposures, with the contact between the Oatman andesite on the southwest and the Gold Road latite on the northeast. The vein fissure is a fault along which the Gold Road latite has been dropped against the Oatman andesite. The maximum length of the United Eastern ore shoot is about 800 feet, and practically all of it lies southeast of the United Eastern No. 2 shaft. Northwest of the shaft the vein decreases in size, and before the long north and west crosscuts shown in Plate X are reached all vein matter has disappeared and only a gouge-filled fissure continues. Where the two crosscuts leave the drift the andesite is cut by a complex of rather irregular and individually nonpersistent fissures with no vein matter. It looks at this point as if the United Eastern vein had practically come to an end. About 1,000 feet farther northwest, however, the Red Cloud workings of the Tom Reed Co. are on a vein which is in line with the United Eastern and is supposed to be the same vein. So far as known, no body of ore has yet been found in the Red Cloud.

GOLD ROAD VEIN.

The Gold Road vein lies about $1\frac{1}{2}$ miles northeast of the Tom Reed vein and has approximately the same strike—nearly northwest. Not only does the vein appear to have originally cropped out conspicuously at many places, but it has been well exposed by a chain of cuts, shafts, and open stopes for nearly a mile southeast of Gold Road. One of the old surface stopes is shown in Plate IX, A, a view taken where the vein is crossed by the Oatman-Kingman road, near the No. 3 shaft of the Gold Road mine. The general dip of the vein is 80°–85° NE. It follows a fault fissure on which there has been a normal displacement, with a throw of probably over 300 feet. This displacement appears to have occurred before and during the deposition of the vein.

A general view showing Gold Road Gulch and the two principal shafts of the Gold Road mine is given in Plate XII, A.

At the time of visit, in 1921, the Gold Road mine was idle, and the only part of the underground workings to which access could be gained was the Line Road tunnel, shown in the longitudinal elevation of figure 7. This elevation, which has been compiled from maps on various scales, is probably not complete, but it indicates the existence of two main ore shoots, one in the vicinity of the No. 1 shaft, near Gold Road, and the other in the vicinity of No. 2 and No. 3 shafts, in Gold Road Gulch. The lode is a fairly regular sheeted zone 100 feet or more in total width. It appears to consist throughout most of the workings of two main nearly parallel veins, known as the North vein and the South vein, accompanied by many smaller stringers. As seen in the Line Road tunnel there is a tendency for stringers to branch off to the south—that is, into the footwall—at a small angle with the general course of the vein. Some of these have been followed in the hope of finding ore, but they are said to die out gradually as they diverge from the main vein. The vein as a rule is unaccompanied by gouge, but it is cut obliquely by a few faults that have shattered the quartz.

The material of the Gold Road vein consists generally of many thin layers of fine-grained flinty or chalcedonic quartz, so that the vein in cross section is beautifully banded, the effect being enhanced by the variety of coloring of the different layers, which range from white, pale yellow, green, or buff to dark brown. In some parts of the vein the deposition of these layers has been undisturbed; in others banded quartz of one generation has been shattered and the fragments cemented by similar quartz of a later period. Calcite occurs in distinct veinlets younger than the quartz and, so far as known, is barren. Some adularia is associated with the quartz but is recognizable only under the microscope. Pyrite is apparently absent. Free gold was not seen in 1921 but is said to be visible occasionally in the ore, which, as noted on page 46, is of low average grade.

Northwest of the No. 1 shaft of the Gold Road mine the vein is traceable for only a short distance over the surface. It was explored in this direction from the West Gold Road shaft, but, so far as known, no ore was found. The geologic map (Pl. I) suggests that the Gold Road vein may represent part of a zone of fissuring that, extending northwestward diagonally across Silver Creek, coincides with the contact between the Gold Road latite and Oatman andesite north of the Sun Dial shaft and is represented by the Gaddis & Perry or Black Wonder lode, north of Hardy Mountain. Proof of this possible continuity, however, is not at present available.

MOSS VEIN.

The Moss vein is one of the most conspicuous veins in the district and was probably the first to be worked. The Moss mine in early days yielded about $250.000 from a body of high-grade ore near the surface, but subsequent exploration of the vein has failed

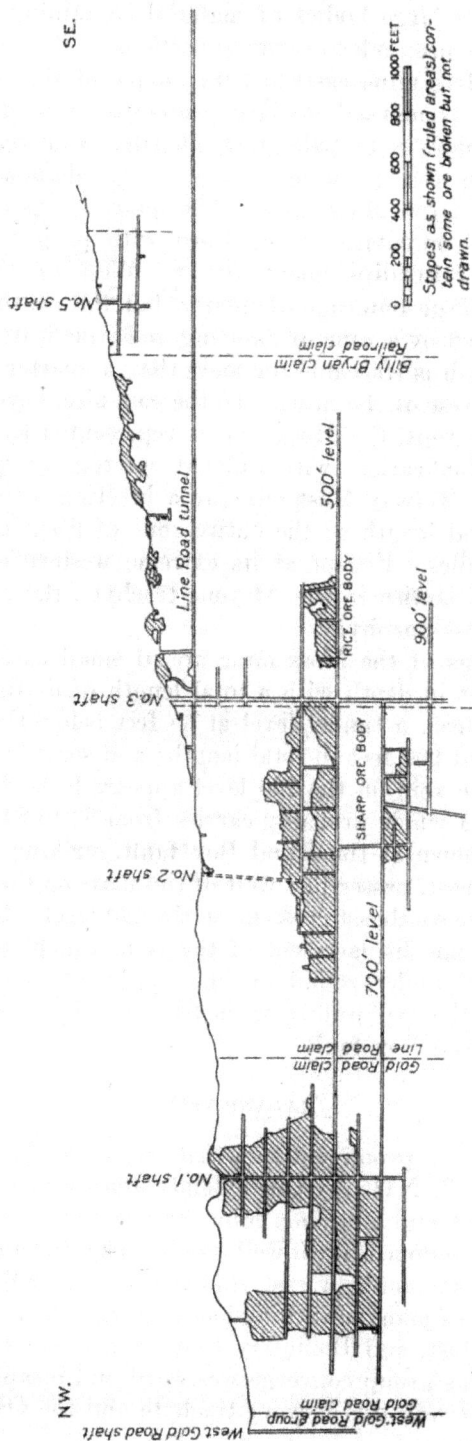

FIGURE 7.—Longitudinal elevation of the Gold Road mine.

to discover any large bodies of material containing sufficient gold to be classed as ore under present conditions.

From the Moss mine eastward the course of the vein, about 12° south of east, is marked by large outcrops of white quartz associated with colorless to pale-green fluorite. Calcite, although not conspicuous in the croppings, is a fairly abundant constituent underground in the Moss mine. The quartz is generally cellular, the little vugs or cavities being lined with projecting crystals of quartz. The vein dips about 70° S. West of the Moss mine there are no large outcrops of quartz, but the continuation of the vein is indicated by a zone of fissuring and silicification in the Moss porphyry, which is traceable for more than a quarter of a mile over the high hill west of the mine. To the east also, beyond the line of prominent outcrops, the Moss vein is represented by a zone of fissuring and silicification, with isolated outcrops of quartz and fluorite, past the Midway Moss mine, to a junction with the Mossback vein. The total length of the entire zone of fissuring and veining is about 2½ miles. Except at its extreme western end, where the zone loses its identity in the Alcyone trachyte, the general country rock is the Moss porphyry.

The workings of the Moss mine are of small extent, comprising a level 220 feet in depth with a total length of drifts and crosscuts of about 750 feet; a tunnel level at 65 feet below the collar of the shaft and about 900 feet in total length; and some irregular surface openings. The vein on the 220 level appears to be fully 90 feet in width and as a whole probably carries from $3 to $4 a ton in gold.

A fissure known as the Blind Boy fault, striking northwest and dipping northeast, passes just west of the shaft on the surface and is followed by the southeast crosscut on the 220 level. This apparently has effected some displacement of the vein, which in 1921 had not been identified underground on the southwest side of the fault. Nearly all of the exploratory work has been done on the northeast side of the Blind Boy fault.

LELAND VEIN.

The Leland is a prominent vein that crops out over the summit of Leland Hill (Pl. XII, *B*) and extends southeastward through the Swiss-American and Golconda mines to a junction with the Pioneer vein at the abandoned Treadwell shaft. The general strike of the vein is about 30° south of east, and the dip 70° SW. The vein is one of a group of many branching lodes, among which are the Pioneer, Lilah, Gold Dust, and Boundary Cone veins. On Leland Hill the members of this group converge westward and disappear at or near the contact between the Alcyone trachyte and the Oatman andesite, on the west slope of the hill. There is evidently some faulting and disturbance of the Oatman andesite where the vein disappears, but the exact manner in which the vein terminates is not clear. That there is nothing in the character of the Alcyone trachyte itself to prevent the continuation of the vein west of Leland Hill is indicated by the fact that the vein is strong in this rock between the Golconda shaft and the Pioneer vein, appearing in places as a broad stringer lode over 50 feet wide.

The Leland vein yielded, according to Schrader, about 4,500 tons of ore in 1903 and 1904 from stopes worked through tunnels in Leland Hill. The vein as seen in these old workings is a stringer lode with no gouge and no definite walls. The stopes are 15 feet in maximum width. The vein material is the usual quartz-calcite aggregate of the Oatman veins.

About 500 feet south of the Leland vein on the surface is the Mitchell vein, which shows croppings of rather promising quartz and calcite. Its strike is nearly parallel with that of the Leland vein, but the dip is north.

The Mitchell shaft, the only deep working on the Leland group, was sunk on this vein to a depth of more than 700 feet. The two veins are supposed to come together a short distance above the 700 level. The exploratory work from this shaft, which appears to have been neither extensive nor well planned, developed no ore, and the 700 level is now under water.

GOLD DUST VEIN.

The Gold Dust or Victor-Virgin vein lies about a mile southwest of the Tom Reed vein, with which it is approximately parallel, and crops out strongly on both sides of Oatman Wash at Old Trails. It is a stringer lode with, as a rule, no well-defined walls and shows considerable variation from place to place. In some exposures it consists of a quartz-calcite vein as much as 12 feet wide accompanied by innumerable small stringers, particularly in the southwest wall, as shown in Plate VII. In other places it is a brecciated mass of andesite, the fragments partly silicified and cemented by quartz and calcite. The country rock is the Oatman andesite.

The vein has been explored for a length of about 500 feet by the workings of the Gold Dust mine. Two ore bodies have been stoped. The larger, southeast of the main or No. 1 shaft, was about 200 feet long and extended from the surface to a depth of about 160 feet. The pay shoot was narrow, the stopes being kept to the smallest practicable working width. A smaller body was stoped between the surface and the first level near the No. 2 shaft, about 450 feet northwest of the main shaft. At the northwest end of the 500 level, as seen in 1921, the vein splits into two or more branches, illustrating the same tendency toward a northwesterly divergence that is shown by the Gold Dust and Boundary Cone·veins at the surface.

PIONEER VEIN.

The Pioneer vein, formerly known as the German-American vein, lies about halfway between the Gold Dust and Leland mines and athwart the general zone of fissuring to which the Leland and Gold Dust veins belong. Its general strike is N. 13° W., and its dip 80°–85° E. The vein is distinctly traceable over the surface for nearly a mile and for much of this distance follows the contact between the Alcyone trachyte on the west and the Oatman andesite on the east. This is clearly a fault contact, the andesite being dropped against the trachyte. How much of the displacement occurred at the original opening of the fissure and how much since the deposition of the vein could not be determined.

The German-American mine, as it was then called, was in operation in 1907, at the time of Schrader's study of the district, and he devotes three and one-half pages to its description. The workings were afterward carried deeper, but in 1921 all work had ceased and no examination of underground conditions was practicable. There are three shafts. The southernmost, the Thirty-fifth Parallel, 220 feet deep, and the Treadwell, next to the north, 340 feet deep, had been dismantled and abandoned prior to 1921. The northernmost shaft, the Pioneer, 420 feet deep, caved while an attempt was being made to unwater it for examination early in 1921. A number of ore shoots were stoped, particularly near the Pioneer shaft. These were as much as 400 feet long and 6 feet wide, and the ore had a maximum tenor of $40 a ton in gold. All the ore bodies were exhausted above the 400 level. According to Mr. George F. Moser, superintendent of the mine, the ore was cut off by a fault which, from his oral description, appeared to be a strike fault with a dip very nearly the same as the vein. A strong fissure filled with gouge, which may be the fault referred to, can be seen accompanying the vein near the collar of the Thirty-fifth Parallel shaft, as shown in Plate VIII, *B*.

MIDNIGHT VEIN.

The Midnight vein, which was worked in the Midnight mine, about 2¼ miles northwest of Oatman, resembles the Pioneer in having a more northerly strike than most of the Oatman veins. Like that vein also, it is accompanied by or fills a fault fissure. The Midnight vein, however, dips west, and the downthrow is in the same direction.

The Midnight mine, which was described by Schrader,[12] has been idle for years, and the workings were not accessible in 1921. The

---

[12] Op. cit., p. 194.

rock on the west or hanging-wall side of the fissure at the mine is the Oatman andesite. On the footwall side, however, are some areas of the peculiar shale that in this part of the district separates the Oatman andesite from the Alcyone trachyte. The occurrence of this shale suggests that the lower workings of the Midnight may be in the Alcyone trachyte.

OTHER VEINS.

There are many other prominent or well-defined veins in the district, such as the New York, Mossback, Gaddis & Perry, Hardy or Record Lode, Pasadena, Blue Bird, Telluride, Lazy Boy, Esperanza, Crown Point, and Highland Chief, which deserve description in a complete report on the district. That they and some others not named are omitted from this preliminary report does not mean that they are necessarily less important or less promising than some that are mentioned.

PRACTICAL CONCLUSIONS.

To men interested in the mining development of the Oatman district, answers to certain questions are of great importance. Some of these questions are as follows: Are the ore deposits confined to any particular rock or rocks? What are the chances of finding additional ore bodies below or at greater depth than those now known? At what depth will the known productive veins be found in pre-Cambrian granite, gneiss, or schist? At what depth will veins now worked in the Oatman andesite go out of that rock, and what will be found below it? Are the large ore bodies confined to any particular area, or may they be reasonably sought for in other parts of the district?

Unfortunately, none of these questions can be positively and definitely answered. The geologist can not see below the surface, and in spite of all the study given to ore deposits, the conditions that determine why a certain vein should contain an ore shoot worth millions of dollars while others are barren remain practically unknown. Ore deposition takes place under circumstances that can not be reproduced in the laboratory. No one knows with certainty where the gold comes from or what delicate balance of complex circumstances determines its deposition in a particular place. We know that there is a connection between the occurrence of most gold ores and igneous activity. It is generally supposed, with considerable evidence in favor of this view, that the solutions which deposited the gold came wholly or in part from solidifying igneous material or magma. But notwithstanding the vigor and ability with which this opinion has been maintained by many geologists, it is certainly very far from being so completely demonstrated as is, for example, the general theory of organic evolution or the dependence of certain diseases upon particular bacteria.

No one who visits Oatman can fail to be impressed with the large number of well-defined veins, consisting chiefly of quartz and calcite. Some crop out boldly and conspicuously; others, although less prominent, can readily be followed over the surface; still others, such as the United Eastern and parts of the Tom Reed vein, although large and productive, give very slight surficial indications of their presence. A notable feature of these veins is the general absence of minerals with metallic luster. The veins, in comparison with those of most other districts, have a barren appearance, and it must be admitted that few of them, on closer investigation, have dispelled this unfavorable first impression. In two lodes only, the Tom Reed and its branches and the Gold Road, have ore bodies large enough to warrant fairly extensive mining operations been found. Good ore has been found in 10 or 12 veins, and some are credited with outputs amounting to a few thousand dollars, or even, as at the Moss mine, to a quarter of a million dollars. These ore bodies, however, have been small and have been exhausted within a disappointingly short distance from the surface.

On the whole, therefore, in spite of the rather extraordinary number of veins in the Oatman district, such evidence as is available indicates that comparatively few of them contain ore shoots that would justify mining and milling equipment to handle over 200 tons of ore a day.

It can not be said that the veins of the Oatman district are confined to any particular rock, although, as shown in Plate I, the more conspicuous ones, those recognizable in the course of geologic mapping, are especially abundant in certain formations and rather lacking in others. They are more numerous, for example, in the Esperanza trachyte, the Oatman andesite, and parts of the Gold Road latite than they appear to be in the Alcyone trachyte. This difference, however, may have no relation to the character of rock but may be simply a matter of areal distribution. For example, if the Esperanza trachyte were removed, the same veins that now show so conspicuously south of Iowa Canyon might appear with equal distinctness in the underlying Alcyone trachyte. Some evidence that this may be the case is furnished by the Leland vein, which between the Golconda and Pioneer mines crops out strongly in the Alcyone trachyte. On the other hand, Schrader's report in many places records the opinion that the Alcyone trachyte (his "older andesite") is decidedly unpromising as regards the possible occurrence of ore, and this view is rather generally held in the district. It is quite true that the strong Leland vein does not appear to continue into the

Alcyone trachyte west of Leland Hill, that this rock as exposed at the surface shows comparatively few veins, and that no ore bodies of any importance have yet been found in it. There is some cause, therefore, for regarding it with disfavor. Nevertheless the rock is not very different in character from the Oatman andesite, and there is no apparent reason why, under favorable conditions, it should not contain ore. That it is everywhere barren or is a rock of such character that ore can not be deposited in it is certainly far from having been proved.

The mines near Oatman, the United Eastern, Tom Reed, Big Jim, and Aztec, are mainly in the Oatman andesite, and the question has naturally arisen, At what depth will the shafts go out of this rock? This can not be answered with precision, for two reasons: First, the attitude or dip of the Oatman andesite as a whole is not known. Second, as the Oatman andesite is known in some localities to break intrusively through the older formations, it can not be assumed that its under surface is approximately plane, like that of a bedded formation such as sandstone. The Mitchell shaft of the Leland mine, for example, is reported to have gone through the bottom of the Oatman andesite into the Alcyone trachyte but to have entered Oatman andesite again at greater depth.

Attention may now be called to Plate I, which shows a section across the district approximately at right angles to the general strike of the formations. It will be observed that the Oatman andesite, where crossed by this section, is bounded on the west by the Pioneer fault. Its actual base at this point is not known but is assumed to be about 400 feet below the surface. If so, and if, moreover, the dip is regular, if the apparent thickness of the andesite between the Pioneer fault and the town of Oatman is not increased or diminished by faulting, if the bottom of the flow is a plane surface, and if the average dip is zero, then the bottom of the Oatman andesite under the town of Oatman should lie at a depth of about 800 feet. The andesite has been ascertained, however, from shafts and diamond-drill holes to be at least 1,500 feet thick vertically at that point; this thickness corresponds to a dip of 5°. If the dip is 10° and the conditions are as assumed, the depth to the bottom of the flow under Oatman would be about 2,200 feet. If, as seems probable, the dip is at least 15°, the depth would be 2,800 feet. A dip of 20°, which is possible, would give a depth of 3,400 feet.

Probably none of the conditions assumed is in accord with fact, but it is impossible to take account of the possible departures from the assumptions made. As a working hypothesis, it may be concluded that it would be necessary to sink from 2,500 to 3,000 feet to penetrate the Oatman andesite under the town of Oatman. The reader is again reminded that this is not an accurate determination, but is merely the best estimate possible under the circumstances.

There has been some speculation among mining men as to what would be found beneath the Oatman andesite, and the possibility of the mines going into pre-Cambrian granitic rock has often been discussed. It is highly probable, however, that the Oatman andesite is underlain by the Alcyone trachyte, which, if it is anything like as thick as it appears to be west of Oatman, would give a vertically measured thickness of at least 2,600 feet of volcanic rock to be penetrated below the bottom of the Oatman andesite, on the supposition of a 15° dip. The granite is not likely to be reached in the mines near Oatman.

Should the shafts or drills near Oatman go through the Oatman andesite, that fact will probably be indicated by the occurrence of distinctly bedded shaly layers such as generally mark the top of the Alcyone trachyte.

The largest and richest ore body yet found in the district, that of the United Eastern mine, extended from a point 270 feet below the surface to a depth of about 1,070 feet, having thus a vertical range of about 800 feet. It was about 1,000 feet in total length and as much as 50 feet wide. It exemplifies what is rather rare—a large ore body which was intact when found, having undergone no erosion, and of which the lower limit is fairly well determined. The more northwesterly of the two ore bodies worked from the Benjamin Harrison shaft of the Tom Reed mine was stoped from a point about 70 feet below the surface to a depth of a little more than 1,300 feet (to the 1,400 level). Its maximum stope length was 650 feet on the 300 level, and the length decreased downward to less than 150 feet on the 1,400 level. The second ore body, which came practically to the surface at the Benjamin Harrison shaft, was stoped to a depth of nearly 900 feet. Its maximum stope length was about 600 feet, on the 500 level. The Aztec-Big Jim ore body, which is probably the longest continuous ore shoot known in the district, has its relations to the surface complicated by the Mallery fault and is not yet fully defined by stoping. It apparently has a stope length in the Aztec and Big Jim workings of fully 1,500 feet. It is doubtful whether any considerable part of this vein will be worth stoping below a depth of 600 feet.

Not much information could be obtained in 1921 concerning the Gold Road ore bodies. The maps seen indicate that the ore body stoped from the No. 1 shaft, near Gold Road, extended from the surface to a depth of 700 feet, with a maximum stope length of 800 feet. According to Schrader,[13] the ore in this part of the mine was as much as 22 feet wide. Whether the Sharp, Rice, and other ore bodies

---

[13] Op. cit., p. 158.

shown on the maps as occurring in the Line Road and Billy Bryan claims are isolated shoots or parts of one large ore shoot is not known. The maps show no stoping in this part of the ground below a depth of 900 feet, although exploration appears to have been carried considerably deeper.

In the United Eastern mine the best ore was found above the 600 level and, as seen in cross section, showed conspicuous banding of the kind illustrated in figure 3. On the 700 level little or no vein material of this character was found. In the Big Jim mine only about 50 feet of the vein on the 600 level contains ore, and the bottom of the ore shoot is probably only a few feet below the level. The 700 level of the Aztec mine, which had merely crosscut the vein at the time of visit, in April, 1921, showed quartz and calcite but no ore. Subsequent development on this level is understood to have been discouraging.

To sum up, certainly very little ore has been found in the Oatman district below a depth of 1,000 feet, and most of the known ore bodies have terminated at depths considerably less than 1,000 feet.

Although in the Oatman ores the gold was deposited in the later part of the period of vein formation, it appears to be of hypogene origin; there is no evidence, so far as could be determined, of downward enrichment.

The general geologic history of the district indicates that at the time the veins were formed the ore shoots could not have been more than about 3,000 feet below the surface. The deposits are distinctly of the Tertiary bonanza type, formed at comparatively shallow depths, as an aftermath of volcanic activity. In such veins the ore shoots are generally confined to a fairly definite zone of vertical range, the parts of the veins above and, more particularly, below this zone being barren. The larger productive mines near Oatman have yielded and are yielding their output from this productive zone, which corresponds to the conditions of temperature and pressure under which the gold was originally deposited by general ascending and probably thermal solutions. Were the veins of different type, such as the older, more deep-seated veins of the western slope of the Sierra Nevada, the search for additional ore shoots at considerably greater depth than those now explored at Oatman might hold out promise of success. A careful consideration of all the evidence indicates, however, that in the Oatman district there is little chance that additional ore bodies will be found at much greater depth than those now known. This does not mean that some ore may not be mined below what are now the deepest stopes, but the hope that some hundreds of feet below such an ore body as that of the United Eastern another comparable body may be found has little to support it.

The occurrence of the Katherine ore body in the pre-Cambrian rocks north of the Oatman district and the fact that this ore, in structure and mineralogy, rather closely resembles that of the United Eastern mine of course suggest the occurrence of similar ore in the pre-Cambrian rocks under the volcanic flows of the Oatman district. The significance of this piece of evidence, however, would be greater if it could be shown that when the Katherine vein was formed the inclosing granite was covered by 8,000 or 10,000 feet of rock that has since been removed by erosion. Although the former presence of such a covering is possible, it does not on the whole appear probable, and the vein itself does not have the character of one formed at great depth.

In conclusion, it may be said that the Oatman district does not hold out inducements for extensive exploration at depths greater than about 1,500 feet, and that the possibility of finding in granite under the lava flows at Oatman ore bodies comparable to those known is practically negligible. For a strong company or a combination of companies to put down a few drill holes below the ground that has given them such rich returns is probably a wise and prudent expenditure of money, but the chances are believed to be decidedly against a successful outcome.

That additional ore bodies remain to be discovered in the Oatman district at moderate depths is probable. The district, as has been shown, presents some rather unusual difficulties to prospecting of the ordinary kind, as some of the largest ore bodies thus far discovered gave little or no indication of their presence at or near the surface. Some of the most conspicuous vein outcrops apparently have nothing of value beneath them, and no criteria have yet been recognized that will enable a prospector to determine from superficial operations whether or not a vein is likely to be productive at a depth of a few hundred feet.

INDEX.

○